CRITICAL INCIDENT STRESS MANAGEMENT IN AVIATION

Critical Incident Stress Management in Aviation

Edited by

JÖRG LEONHARDT AND JOACHIM VOGT

ASHGATE

© Jörg Leonhardt and Joachim Vogt 2006

All rights reserved. No part of this publication may be reproduced, stored in a retrieval system or transmitted in any form or by any means, electronic, mechanical, photocopying, recording or otherwise without the prior permission of the publisher.

Jörg Leonhardt and Joachim Vogt have asserted their right under the Copyright, Designs and Patents Act, 1988, to be identified as the editors of this work.

Published by
Ashgate Publishing Limited
Gower House
Croft Road
Aldershot
Hampshire GU11 3HR
England

Ashgate Publishing Company
Suite 420
101 Cherry Street
Burlington, VT 05401-4405
USA

Ashgate website: http://www.ashgate.com

British Library Cataloguing in Publication Data
Critical incident stress management in aviation
  1. Aeronautics, Commercial - Employees - Job stress
  2. Stress management 3. Aviation psychology
  I. Leonhardt, Jörg II. Vogt, Joachim
  158.7'2

Library of Congress Cataloging-in-Publication Data
Critical incident stress management in aviation / edited by Jörg Leonhardt and Joachim Vogt.
    p. cm.
  Includes bibliographical references and index.
  ISBN-13: 978-0-7546-4738-6
  ISBN-10: 0-7546-4738-2
 1. Aviation psychology. 2. Aircraft industry--Employees--Health and hygiene. 3. Stress management. 4. Crisis management. I. Leonhardt, Jörg. II. Vogt, Joachim.

  RC1085.C75 2006
  155.9'65--dc22

2006021143

ISBN-13: 978 0 7546 4738 6
ISBN-10: 0 7546 4738 2

Printed and bound in Great Britain by MPG Books Ltd. Bodmin, Cornwall.

Contents

List of Figures		*vii*
List of Tables		*ix*
List of Contributors		*xi*
List of Abbreviations		*xiii*
1	Introduction *Joachim Vogt and Jörg Leonhardt*	1
2	Importance of CISM in Modern Air Traffic Management (ATM) *Ralph Riedle*	5
3	Critical Incident Stress Management in Aviation: A Strategic Approach *Jeffrey T. Mitchell*	13
4	Critical Incident, Critical Incident Stress, Post Traumatic Stress Disorder – Definitions and Underlying Neurobiological Processes *Jörg Leonhardt and Joachim Vogt*	43
5	Critical Incident Stress Management Intervention Methods *Jörg Leonhardt*	53
6	A Brief History of Crisis Intervention Support Services in Aviation *Jeffrey T. Mitchell*	65
7	Critical Incident Stress Management in Air Traffic Control (ATC) *Jörg Leonhardt*	81
8	Critical Incident Stress Management in Airlines *Johanna O'Flaherty*	93
9	Critical Incident Stress Management at Frankfurt Airport (CISM Team FRAPORT) *Walter Gaber and Annette Drozd*	109
10	Case Study Lake Constance (Überlingen) *Jörg Leonhardt, Carol Minder, Sabine Zimmermann, Ralf Mersmann and Ralf Schultze*	125

11	Training in Critical Incident Stress Management – Principles and Standards *Victor Welzant and Amie Kolos*	145
12	Cost Benefit Analysis of a Critical Incident Stress Management Program *Joachim Vogt and Stefan Pennig*	153

Appendices — 171

Index — *177*

List of Figures

1.1	Structure of the book	2
4.1	Different critical incidents (CIs) meet person A in a good and person B in a limited psychophysical condition	45
5.1	Spectrum of care: CISM as the first element in psychosocial help (modified according to Mitchell and Everly 2002); EAP: Employee Assistance Program	54
5.2	Three phases of defusing (modified according to Mitchell and Everly 2002)	58
5.3	Seven phases of debriefing (modified according to Mitchell and Everly 2002)	59
5.4	Know where you are in Critical Incident Stress Debriefing (CISD) (modified according to Mitchell and Everly 2002)	60
9.1	Frankfurt Airport traffic data	110
9.2	Training modules for Crisis Intervention Team	119
9.3	Responsibilities of the FRAPORT Crisis Intervention Team (CIT), CISM Team, medical services, and the joint Air Traffic Control – Airport (ATC-AP) Team	122
9.4	Structure of the CISM Team Aviation (Air Traffic Control – ATC-AP)	123
10.1	Seven phases of debriefing (modified according to Mitchell and Everly 2002)	129
10.2	Introduction phase	129
10.3	Fact phase	131
10.4	Thought phase	133
10.5	Reaction phase	135
10.6	Symptom phase	137
10.7	Teaching phase	138
10.8	Re-entry phase	140
12.1	The evaluation model of the feasibility study	155
12.2	Model of expected benefits of CISM in an air navigation service provider (ANSP; SV: supervisor; COS: chief of section)	158
12.3	Self-management of air traffic controllers as process model (individual level: HR: human resource; HF: human	

	factor; T: training; TRM: team resource management)	161
12.4	Self-reported performance in the week before, during, and after a critical incident (CI) or controllers requesting or rejecting CISM	162
12.5	Operative leadership as a process model (process level; HR: human resource; HF: human factor; T: training; TRM team resource management)	164
12.6	Strategic steering as process model (strategic level)	166

List of Tables

3.1	Psychotherapy and crisis intervention	23
4.1	The Selye (1936, 1950) distress and the Cannon (1914) eu-stress axis in an integrative model (modified according to Henry and Stephens 1977, p. 119)	51
8.1	Emergency levels within the airline industry	96
8.2	Critical incident stress debriefing criteria formulated between American Airlines Employee Assistance Program counselors and the flight attendants union Association Professional Flight Attendants	101
9.1	Examples for missions of the FRAPORT Medical Team providing counseling for victims of accidents and/or their relatives	119

List of Contributors

Annette Drozd is physician (occupational and internal medicine) and medical doctor (Dr. med.) at Frankfurt Airport (FRAPORT). a.drozd@fraport.de

Walter Gaber is physician and medical doctor (Dr. med.). He is Vice President Human Resources and Medical Director at Frankfurt Airport (FRAPORT) as well as an international consultant in disaster management. w.gaber@fraport.de. www.eagosh.com

Amie Kolos BA (cand.) is currently completing undergraduate training in Psychology at Towson University, Towson MD, USA. She is affiliated to the International Critical Incident Stress Foundation, in Ellicott City MD, USA in the training and education division. www.icisf.org

Jörg Leonhardt holds a Masters degree in Social Work (MSW). He is a family therapist, CISM trainer (ICISF), and is currently working as a human factors expert in the Safety Management Division of the German air navigation service provider (ANSP) Deutsche Flugsicherung (DFS). He developed the DFS CISM program and has conducted peer training for several ANSPs throughout Europe. joerg.leonhardt@dfs.de

Ralf Mersmann is air traffic controller at Karlsruhe upper area control center, an ICISF trained peer, and a member of the DFS crisis intervention team. ralf.mersmann@dfs.de

Carol Minder is air traffic controller at Zürich area control center and ICISF trained peer. carol.minder@skyguide.ch

Jeffrey T. Mitchell, Ph.D., CTS, is Clinical Associate Professor, Emergency Health Services at the University of Maryland. He is President Emeritus of the International Critical Incident Stress Foundation, which he founded together with George S. Everly. jmitch@umbc.edu. www.icisf.org

Johanna O'Flaherty holds a doctorate degree in Clinical Psychology from Pacifica Graduate Institute, California. She developed and implemented Crisis Response Programs for TWA and served on a focus group with the Airline Transport Association (ATA) to develop industry guidelines for Crisis Response Teams. For over 20 years, she developed and managed Airline Employee Assistance Programs and joined Sierra Tucson in November 2005. johbowler@earthlink.net

Stefan Pennig holds a diploma in psychology (Dipl.-Psych.) and economics (Dipl.-Kfm.). He has worked as a consultant in the areas performance and leadership for 15 years. He developed the Human Resources Performance Model for the cost-benefit-analysis of CISM and other human factors, human resources, and training programs. pennig@cneweb.de. www.context-online.de

Ralph Riedle was air traffic controller and is now Managing Director Operations of DFS responsible for nearly 1800 operational employees. He took the management decision to develop a CISM program at DFS. ralph.riedle@dfs.de

Ralf Schultze is air traffic controller at Karlsruhe upper area control center, ICISF trained peer, and member of the DFS crisis intervention team. ralf.schultze@dfs.de

Joachim Vogt holds a doctorate (Dr. phil.) and a professorial degree (PD, Habilitation) in Psychology. He is currently Associate Professor for Work and Organizational Psychology at Copenhagen University. He received CISM basic and advanced training. His current research focus is the cost-benefit-analysis of CISM and other human factors, human resources, and training programs. joachim.vogt@psy.ku.dk

Victor Welzant holds a doctorate degree in Psychology (Psy. D.). He serves as the Director of Education and Training and faculty member of the International Critical Incident Stress Foundation in Ellicott City MD, USA. Victor's professional activities include serving as a member of the Maryland Disaster Behavioral Health Advisory Committee, as well as a consultant to the Maryland Mental Hygiene Administration on disaster behavioral health practice and training. He also directs the Acute Trauma Services program at Sheppard Pratt Health System in Towson, MD,USA. welzant@icisf.org. www.icisf.org

Sabine Zimmermann is air traffic controller at Zürich area control center and ICISF trained peer. sabine.zimmermann@skyguide.ch

List of Abbreviations

ACC	Area Control Center
AFA	Airline Flight-attendants Association
ALPA	Air Line Pilots Association
AOC	Airport Operational Committee
APA	American Psychological Association
APFA	Association Professional Flight Attendants
ANSP	Air Navigation Service Provider
ATC	Air Traffic Control
ATC-AP	Air Traffic Control – Airport CISM Team
ATCO	Air Traffic Controller
ATM	Air Traffic Management
CBT	Cognitive Behavioural Therapy
CI	Critical Incident
CIRP	Critical Incident Response Programs
CISD	Critical Incident Stress Debriefing
CIS/R	Critical Incident Stress Reactions
CISM	Critical Incident Stress Management
CIT	Crisis Intervention Team
CMB	Crisis Management Briefing
COST	Certificate of Specialized Training
CRM	Crew Resource Management
DFS	Deutsche Flugsicherung (German ANSP)
DLH	Deutsche Lufthansa
EAP	Employee Assistance Program/Professional
EMDR	Eye Movement Desensitization and Reprocessing
ERIC	Emergency Response and Information Center
FAA	Federal Aviation Administration (of the USA)
FRAPORT	Frankfurt Airport Group
HPA	Hypothalamic Pituitary Adrenal axis
HPM	Human Resource Performance Model
ICD	International Classification of Diseases
ICISF	International Critical Incident Stress Foundation
IFATCA	International Federation of Air Traffic Controller Associations
MHP	Mental Health Professional
NASA	National Aeronautical and Space Administration (of the USA)
NATCA	National Air Traffic Controllers Association (of the USA)
NTSB	National Transportation Safety Board
PTSD	Post Traumatic Stress Disorder

ROI	Return on Investment
SAFER-R	Stabilize, Acknowledge, Facilitate, Encourage, Recovery, Referral
TFT	Thought Field Therapy
TIR	Trauma Incident Reduction
TRM	Team Resource Management
TWA	Trans World Airlines
UAC	Upper Area Control
WHO	World Health Organization

Chapter 1
Introduction

Joachim Vogt and Jörg Leonhardt

The idea of this book was born some years ago. As human factors practitioner and researcher we noticed an increasing need and interest in Critical Incident Stress Management (CISM) in aviation. However, available CISM books were mostly related to fire fighters and emergency rescue personnel. Thus, we started gathering experts with decades of experience in applying and researching CISM in aviation.

The protagonist organizations in aviation are airports, airlines, and air traffic control with close cooperation within and between them (Figure 1.1). Therefore, the book is arranged around these organizations' views on crisis management: Special organizational requirements, implementation and structure of CISM, rules and procedures, advantages, benefits and experiences in general are outlined by a representative of the respective protagonist. For the airline and airport perspective we won over Johanna O'Flaherty, who is a consultant for American Airlines, and Walter Gaber, the Director of Medical Services at Frankfurt Airport, together with his colleague A. Drozd for Chapters eight and nine, respectively. The air traffic control perspective in Chapter seven is given by Jörg Leonhardt, CISM program manager at the German air traffic control service provider Deutsche Flugsicherung DFS. Chapter ten is a case study of CISM application after a major aviation accident – the mid air collision over Lake Constance. The case study gives a tripartite perspective: The debriefing process in theory, the perspective of peers involved and the perception of participants. This facilitates a look into a CISM process and the reader can reflect on all three levels how a debriefing can develop its quality. Many thanks are due to Carol Minder and Sabine Zimmermann, affiliated to the Swiss air traffic control service provider skyguide, for sharing their perception as participants, and to Ralf Mersmann and Ralf Schultze, DFS, for providing the peers' view.

Experience did prove that the success of a CISM program depends on the qualification of the peers and the support and understanding of the management. Principles and standards in the training of peers are described by Victor Welzant, International Critical Incident Stress Foundation ICISF, in Chapter 11. The management perspective is given in Chapter two by Ralph Riedle, Managing Director Operations of DFS responsible for nearly 1800 operational employees. Before Ralph joined DFS management he has worked as an air traffic controller and thus knows the requirements and limitations of this work very well. Certainly, his contribution will be a significant inspiration to other aviation top managers in acknowledging and implementing CISM in their organizations. The empirical evidence that CISM significantly supports the strategic goals of an aviation organization with many intangible and also monetary benefits is given in Chapter 12 by Joachim Vogt, associate professor for work and organizational psychology at Copenhagen University, Denmark, and Stefan Pennig, context Leadership. Performance.Consulting, Essen, Germany. Chapter 12 reports the feasibility study and the main investigation of CISM effectiveness and goal attainment which were conducted at DFS on behalf of EUROCONTROL, Brussels, and the German Federal Institute of Occupational Safety and Health.

The chapters introduced so far focus on practical issues of CISM implementation, operation, and evaluation. The definition of important CISM terms, basic theoretical and methodological background is given in Chapters 4 and 5.

Finally, we are happy and proud that we could persuade Jeffrey T. Mitchell, clinical associate professor at the University of Maryland and President Emeritus of ICISF, to contribute Chapters 3 and 6. His knowledge and large-scale experience as the inventor of CISM are paramount for this book. Chapter 3 outlines the basis for a comprehensive and clear understanding of CISM. Chapter 6 describes the history of CISM in aviation from the perspective of Jeff's first-hand knowledge and experience. The history of CISM in aviation may be short in time, but it is manifold and intensive.

We are grateful to all authors of this book for their dedicated writing and patience in going through several revisions with us. Thanks are also due to Gillian Farrell for improving the language in Chapters four, five, and seven, to Maren Maziul who formatted the typescript, and to Ingo Lahl and Dirk Haape, who designed an emblem for the DFS peers and made it available for the book cover. Finally, we would like to acknowledge the pleasant and effective cooperation with Guy Loft and Sarah Horsley, our editors at Ashgate.

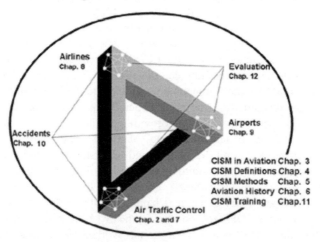

Figure 1.1   Structure of the book

Figure 1.1 summarizes the structure of the book. It shows several airlines, airports, and air traffic control service providers cooperating with each other and using CISM, as described in Chapters two, and seven to nine. Natural catastrophes, terror attacks, or accidents – like the one described in Chapter ten – can affect all aviation organizations. CISM is one important preparation for this. Chapters 3 to 6 and 11 provide general information on CISM and are displayed as the shielding surrounding of all aviation organizations. Evaluation and improvement of the respective organizations CISM program has to take place mainly within the organization, aligned with its specific strategic goals (Chapter 12).

The chapters are interwoven and independent at the same time. The reader will find some redundancies and repetitions; these are intended because some reader groups will probably concentrate on the chapters relevant to their organization only.

Mario Lucci, an air traffic controller attending a CISM training session coined the phrase 'set heading to the positive'. We feel that this phrase precisely expresses the main goal of CISM, namely to start regaining balance after critical incidents which are the unfortunate side effects of increasing air traffic. International aviation has acknowledged this and the number of organizations requesting peer trainings and consultation in developing CISM programs is increasing – hundreds of air traffic controllers worldwide went through a peer training and the network is growing. We hope that this positive development will continue and that this book can make a decent contribution.

Copenhagen and Darmstadt, May 2006

Chapter 2

Importance of CISM in Modern Air Traffic Management (ATM)

Ralph Riedle

Modern ATM, Future Requirements, and the Role of CISM

Management by objectives, flat hierarchies, consistent delegation and teamwork are the basic principles of modern management, particularly in the field of air traffic control. The ability to work in a team does not only refer to a team of managers but also to the way managers cooperate with their staff. Teamwork can only be successful if there is open and efficient communication.

Flat hierarchy means that there are only a few management levels which, in turn, means that everyone has to bear more responsibility. Managers delegate more tasks to their staff, but they must ensure that they have the adequate responsibility and competence. Tasks should always be delegated to the lowest level, for example tasks and responsibilities which can, and should, be performed directly by a supervisor must be delegated to him and not to the head of branch. In this context, competence also comprises the authority to make decisions concerning the task.

Delegation enhances the efficiency of the management process and ensures that all staff members are responsible for achieving the company's objectives. The basic principle that the person to whom the task is delegated has the necessary technical and personal competence and authority to make the decisions required for the delegated task, must always be adhered to.

Safety is the paramount objective of air traffic control with such a management principle. It must be ensured at all times. In addition to safety, however, the management must also focus on enhancing cost-effectiveness and productivity of the organization. This may result in conflicts which modern management has to deal with. If cost-effectiveness and productivity are not taken into consideration, safety could be used to justify any decision. Safety, as the most important principle and corporate objective, must not be compromised; nevertheless cost-effectiveness and productivity may not be disregarded. Before a decision is taken, it must first be checked how it will affect safety. Cost-effectiveness and productivity should then be considered in a second step.

In the past few decades, different organizations have undergone essential but different changes. More than ten years ago, the former German Federal Administration of Air Navigation Services was reorganized into a corporate organization under private law. As the next step, a process-oriented organization with different business areas was developed, which were then structured as business units, corporate development centers and corporate service centers. At the same time, increases in fuel (kerosene) price, higher costs for procurement and maintenance of aircraft and lower ticket prices have been putting more and more pressure on service providers and the organizations providing the infrastructure (namely the air navigation service providers, ANSPs) to not only ensure safety, but to ensure safety whilst working cost-effectively.

Today, there are even attempts to base ATC charges on decreasing unit costs.

Some airlines are demanding to pay lower ATC charges in the case of performance shortfalls (for example delays). This would mean that ANSPs would have to grant discounts in the case of alleged performance shortfalls caused by them. This would be a critical development. Companies would be tempted to render their services in a way that would prevent discounts and in a business such as ours, this could compromise safety. Safety would no longer have priority.

An example for this is an airport with a coordinated capacity at which more aircraft than permitted wish to arrive or depart during a certain time period. To prevent performance shortfalls (that is delays for the airlines), minimum separation would have to be infringed in order to meet customer demand. And this would surely compromise safety.

Consultation processes and partnership agreements, i.e. agreements concerning the service levels, with customers are another modern management method. The aforementioned management method comprising the implicit principle of delegation of responsibility and competence is necessary to control such processes.

The demands on controllers have grown in line with the traffic volume, and the increase in workload cannot always be offset by new technologies. Air traffic control has always been marked by teamwork and the demands are still met by teams. Because of the increase in traffic volume, it will not be sufficient to install new technologies. It will also be necessary to enhance team structures and launch optimization programs.

Team Resource Management (TRM) is an essential component and development program for establishing a team structure in which the team members can work without conflicts and in harmony. Modern ATC management must further develop teamwork structures and the controllers' ability to work in teams. TRM is used to reduce the workload of the individual team members and thus increases overall capacity. Controllers are conservative in a positive sense because their business is to ensure safety, and they often feel that changes may jeopardize safety.

Critical Incident Stress Management (CISM) perfectly complements human factor concepts, since it not only helps dealing with a person's individual situation but also affects the way a team works. CISM alleviates the fear of having failed or having contributed to an incident or accident.

In the history of air traffic control, there have been attempts to automate air traffic control and replace controllers with machines or technical systems. When I passed the aptitude test for controllers at the Federal Administration of Air Navigation Services in 1971, I was welcomed with the following sentences:

> We regret that you have chosen a profession which will be obsolete in a few years. Soon, air traffic controllers will be replaced by computers. We are currently implementing partially automated radar control.

We know that this has not happened. Over time and in line with the developments, it was determined that the system has to focus on the person and that the technical systems, procedures and management should support the air traffic controller in handling the increase in traffic volume.

The shift from full automation to systems, technology and procedures which only serve to support the controller, relieve him from routine tasks and enhance capacity, has taken a long time. In today's modern ATC management, technology is considered as a tool for the controller. Similar changes of paradigm have taken place in the entire aviation industry.

Implementation of the DFS CISM Program: Overcoming Prejudices

When CISM was first discussed at DFS and the Board of Managing Directors were informed about it, I learnt about the benefits of CISM over TRM. We were just implementing TRM when CISM was first introduced. I tried to figure out how the systems differ and why we needed a supplementary system. Then I realized that CISM is not an add-on, but an independent system which should be taken into consideration. At the beginning when the structures and the program were not yet defined, I was afraid that such a support and counseling system might require a large number of employees trained in CISM and that such peers would no longer be available to do their regular work.

I was also concerned that it might be a self-fulfilling prophecy and that every controller would need a peer for every little problem so that there would be more counseling than air traffic control. And I feared that the counseling might evolve into lamentations and that controllers would lose their motivation and trust in themselves, their capabilities and competence, since too much talk about problems often leads to a loss of confidence. This happens a lot in an area where the staff have to make decisions and find a solution very fast.

Like many others who are familiar with ATC, I also believed that a controller should be able to deal with a loss of separation himself, since that is part of his job. Today I know that this meant that controllers did not have any support in dealing with their self-doubts and often acted like nothing happened just to appear professional. "I must not show any weaknesses because I am a controller" might have been a frequent thought.

All of my fears and concerns were gradually dispelled.

The more details I was given about the program and its implementation in the field of air traffic control, the more I realized that it was necessary and sensible to support its implementation at DFS. We now have findings about the positive effect of the program on the staff members and about its acceptance within the company (see Chapter 12). A key success factor for the CISM program is ensuring that the person responsible for the implementation shows understanding and can put things into perspective.

CISM can only be implemented successfully, if it is done so within an adequate framework. When I realized how important such a program is for a company with the responsibility and size of DFS, my concerns were dispelled.

Today I know that CISM is very helpful and supportive but does not infringe the limits of responsibility of any of our staff members. The peers who are working in the program must be aware of their responsibilities and limits and offer CISM when there has been a critical incident. This requires sound, well-balanced and excellent training of controllers to become CISM peers because CISM is not the same as general life counseling.

Effects of CISM on Corporate Culture

Of course, CISM has affected the corporate culture in our company; there have been many positive effects with regard to communication and the understanding of reactions to critical incidents. Even though the air traffic controllers were rather skeptical at first, the positive experiences they had and reports they heard led to a great degree of acceptance. When I talked to the managers and operational staff in the branches, I only had positive responses. I was often told by staff members that they considered it important that DFS had implemented and maintained such a program. They honored the support of the management and understood that CISM was implemented to support operational staff.

There were, however, objections when we wanted to conduct an evaluation and draw up annual statistics. However, annual statistics are now standard procedure and they are widely accepted. Our staff realized that the statistics serve the program and are not used against individual staff members. They validate the effectiveness of CISM at DFS.

The importance of the program itself, the commitment of the management for the implementation and maintenance of such a program and the understanding of critical incident stress reactions have ensured that CISM is not only widely accepted, but applied and considered to be helpful. Our staff members feel that it is in place to directly support them.

A good knowledge of what CISM is about and the conviction that such a program is essential determines its status within the company. If the management of a company understands the importance of the program at a very early stage and supports its implementation, the staff will respond positively and see that they will benefit from it.

At a later stage, we realized that the CISM program not only helped our staff members but even increased the cost-effectiveness of our company (see Chapter

12). There is less absenteeism, staff members return to work faster after an incident and the risk that the reactions to a critical incident cannot be controlled, has been mitigated. There have been far less absences and illnesses; even occupational invalidity caused by a critical incident has nearly been reduced to zero.

Both staff members and the company itself benefit from this program. The benefits have to be presented within the company in the form of statistics, evaluations, cost-benefit analyses or at meetings. Staff members should be informed about the program, its use and its benefits. This is the easiest and best way to promote CISM and increase its acceptance.

The cost-effectiveness of CISM can be validated by means of internal statistics and cost-benefit analyses (see Chapter 12).

At DFS, CISM is also used for recording infringements of separation. Every incident report includes information about whether CISM was actively offered and accepted. This is very important since it shows that we, as a company, are serious about CISM and want our staff members to receive adequate and professional counseling.

CISM also affects our safety culture. After the controllers were afraid of being ridiculed if they used CISM in the beginning, it is now considered to be normal and standard procedure. It has become part of the professional understanding of controllers and is completely integrated into their everyday work.

This affects our safety culture in different ways: Mistakes are dealt with more openly, which leads to a *lessons learnt* effect. The know-how that is gained during and after incidents is applied within the company.

Talking about difficult situations often means that the persons involved learn to deal with themselves and the critical situations more openly. This process directly contributes to a safety culture, it helps a person to learn from mistakes and avoid making the same ones again.

It also affects other groups within the organization, such as the safety boards and incident investigators. The open communication improves the safety culture and situation of a company. A culture of trust must be based on the knowledge that a company wants to learn from incidents and not blame or punish the person involved. CISM supports such an approach.

Managers in operations are an essential part of the CISM program. They have to pave the way and ensure that CISM is offered. They support the CISM process because they know about its positive effects and its contribution to the safety culture.

But the most important factor is that the management knows how important it is for the staff that such a program is in place.

Our staff and our management have been shown that learning from mistakes is taken seriously, that dealing with incidents increases safety and that we do not aim at blaming or punishing individual employees. CISM has helped us achieving all this and make it part of our corporate culture.

Recommendations to Other Aviation Managers

The CISM concept was completely new to us and it was difficult to get a feeling for it. My recommendation to my fellow managers is to contact the International Critical Incident Stress Foundation (ICISF) and their European Office. They can provide information and explain the scope and the concept.

Establishing human factor programs is not part of the daily work of managers. They have to consider all abstract and practical aspects in order to determine the dimensions, effects and opportunities. If a similar program exists in comparable industries or companies, the managers should ask for all relevant information, look at the program and benefit from the experiences others made. This helps preventing mistakes and enhances the strengths of the program.

Once all information and reports have been collected, the managers have to decide how the CISM program should be established to ensure that it will be successful. It is very important that a person, who either has or will acquire the necessary abilities, is appointed as program manager and bears the responsibility for the overall project. Such a program manager who takes the company and its objectives into consideration will be able to establish a program that is tailored to the needs of the organization.

Certain features will have to be adapted to the relevant company. An airport has to focus on different aspects than an air navigation service provider or an airline (see Chapters 7, 8 and 9). For the program to be accepted by the staff, however, the program manager must be supported by the management.

The faster the program is accepted by the staff members, for whom it was established, the faster it will be a success.

Permanent success is an important aspect. The management must know the program, the status of its implementation and its value and inform the staff about it. This can be done with figures and statistics or with a strong presence of the management.

The ongoing development of CISM, while sustaining its high quality, also plays an important role.

Continuous information and development characterize the DFS CISM program. Examples are the local exchange of information and lessons learnt, the annual CISM Forum at the DFS headquarters, and the exchange of information by the staff responsible for the program. The value of the program can be upheld by means of continuous investments and the aforementioned meetings and events.

The training of peers (see Chapter 11), the continuous development of the program and the repetition of the information is very important for everyone, but particularly for new staff members and managers. The program must be adjusted and adapted to new challenges and situations. Refresher courses and information campaigns have to be organized on a regular basis. The management has to confirm its support repeatedly and firmly to maintain trust and belief in the program.

Staff members are motivated by their tasks. CISM greatly enhances this motivation because the staff members know that the company provides such programs to enable them to deal with critical situations. The peers deserve acknowledgements as they are very committed, provide services in addition to their regular work and spend time helping their colleagues on a voluntary basis and without additional remuneration. This work should not be made more attractive by higher salaries. It should remain a voluntary job but the commitment should be acknowledged, i.e. in the form of praise or special events for the volunteers.

This acknowledgement can also be shown by facilitating the exchange with other organizations or similar companies. If many staff members participate in such a program, they will get an insight into other corporate cultures, procedures and organizations.

All rational arguments have been described. But it is also very important to feel that the program is good for the company and for the staff members. Critical incidents are always very difficult to deal with for air traffic controllers. I can only say that I am very glad that I was able to contribute to the establishment of CISM at DFS. The feedback I have been receiving shows me that I have done the right thing. And that is all I can do as a manager.

Chapter 3

Critical Incident Stress Management in Aviation: A Strategic Approach

Jeffrey T. Mitchell

Introduction

People who live at great distances from appalling natural disasters like floods, earthquakes, and tsunamis may experience emotional distress and great sympathy for the victims. Typically their distress diminishes in a short time and they return to normal life. Rarely do people express long-term anger and rage about a natural disaster.

Aviation disasters, on the other hand, may trigger, at least initially, intense levels of stress and alarm that later turn into anger and rage against the builders, operators, employees and agencies that control aviation. Expressions of anger, often stimulated by news stories and reports on the accident investigation, may continue for many months, if not years.

It may be mentally easier for people to shake off natural disasters in strange, faraway places with populations who do not speak the same language and whose customs are unfamiliar or even peculiar to them. Perhaps people acknowledge that powerful, unstable, and uncontrollable forces have always existed in nature and have caused disasters throughout history. It is expected, therefore, that natural disasters will eventually occur. In fact, entire communities are encouraged to prepare for them.

The expectations in aviation are much different. No doubt, the industry itself and emergency services organizations prepare for aviation disasters. The average public, however, does not make such preparations. The public assumes that aircraft are carefully designed, built, and tested. They know that there are government agencies that assure safety and establish operating procedures that are supposed to keep crashes from occurring. For the most part, all the testing, regulating, and safety monitoring does what it has been designed to do – prevent aviation disasters.

In actuality, there are very few crashes when considered in proportion to the total number of annual air flights. Since the number of air crashes is low, sophisticated and complex aviation technology is more likely to be trusted. At times, that trust may be irrational. For example, some people make assumptions that air crashes are a thing of the past. Aviation disasters, people surmise, should simply not occur in our modern world. Perhaps that thinking is one way in which people deny their own mortality and attempt to reduce their anxiety regarding flying.

An airplane is something quite familiar to people in well developed countries. Even in less developed countries, airplanes transport vital cargoes and serve as an economic and social lifeline. We may periodically fly in planes or, at least, we know people who do. They bring to us or take away from us our family members, friends or business associates. Each day planes pass overhead and we hear them. They are visible both in the sky and at airports. When a plane crashes, the media floods the site and gathers dramatic and horrific images. Fire, smoke, victims, grieving family members, broken aircraft parts and bits of blackened, twisted metal fill our television screens (Duffy 1979; Freeman 1979; Forstenzer 1980).

An aircraft disaster reduces our ability to deny our own vulnerability. It shocks us into the reality that we too could meet the same fate as those on board the plane. Or, it might be someone we love who could be a casualty. Just the thought of being a potential victim or knowing a victim is terrifying.

An aircraft crash causes even greater emotional shock for people who work within the aviation industry. They feel the same shock, horror and grief as everyone else. However, they also feel responsibility and guilt because they design, manufacture, maintain, operate, communicate with, and control the aircraft that crisscross the sky. The media, politicians, the public and, sometimes, even the airline's corporate leadership are quick to place blame on the employees of an airline and hold them accountable. This often occurs before the actual facts are known (Frederick 1981).

An air disaster can throw people in the aviation industry into a state of emotional crisis. Gate agents have a hard time facing the flying public. Pilots and flight attendants may not wish to fly on board the same type of aircraft as the one that crashed especially if the type of aircraft has a history of several crashes. Ground and maintenance crews and air traffic controllers review their procedures to see if they may be at fault. Few within the industry feel at ease with their work in the months after a catastrophic incident.

It would be wrong to suggest that crisis reactions only occur as a result of devastating aircraft disasters. Emotionally distressing situations on a much smaller scale occur daily throughout the industry. Passengers get injured or sick while flying and some even die on board aircraft. Most flight attendants will agree that child-related severe illness or injuries are among the most disturbing incidents they encounter. In addition, there are the *close calls*, severe turbulence, threatening or violent passengers, equipment failures, hard landings, and small accidents. The media rarely covers those situations but personnel are still left shaken and distressed.

It is easy to forget that aviation employees are people first. They have families, children, and lives outside the job. They encounter threats and losses in their personal lives that could potentially impair their performance on the job. They might come to work and have to leave a sick child at home. They have the usual conflicts of family life. They suffer divorce, financial distress and a host of frustrations. They might just need someone to listen to them and reassure them so that they could work effectively.

The aviation industry, at all levels, needs quality crisis support programs to assist its employees and keep them functioning at peak performance levels. That means peer and professional crisis responders must be properly trained and organized to respond quickly to an individual or group crisis and to provide the right support services at the right time and under the right circumstances. A strategic approach to crisis intervention is a necessity, not a luxury, for all branches of aviation. These programs have proven their value time and again in small and large-scale tragedies on many air carriers and in many different types of operations within the aviation industry.

Caution: *Reading the contents of this chapter, or even the entire book, is not substitute for adequate crisis intervention training and supervised practice. Furthermore, no one should construe the guidelines in this chapter as if they were clinical prescriptions or a substitute for professional mental health services. When in doubt contact a mental health professional.*

The Nature of Crisis Intervention and Critical Incident Stress Management

We can only engage in effective crisis support services for the many branches of the aviation industry if we have a sufficient knowledge of the basic definitions and core principles of crisis intervention. This section will cover some important definitions.

Critical Incidents

Critical incidents are emotionally powerful events that overwhelm an individual's or a crew's ability to function normally (see Chapter 4). They are the starting point for a crisis reaction. Critical incidents are perceived to be overwhelming, threatening, frightening, disgusting, dangerous, or grotesque circumstances. One or more of the following disturbing situations can throw a person working in any aspect of the aviation industry into a heightened state of emotional turbulence known as a *crisis*:

- Accidents
- Disasters
- Illness
- Deaths
- Violence
- Threats of violence
- Situations posing danger to one's life or safety
- Unwanted pregnancies
- Alcohol or drug overdoses
- Financial losses
- Public humiliations
- Personal rejections
- Diseases

- Property losses.

This list represents but a few of the distressing critical incidents that may occur in the course of a person's life (Caplan 1961, 1964, 1969).

Crisis

A crisis is defined as an acute emotional reaction to some powerful stimulus (critical incident) or demand in a person's life or in the life of one of their loved ones. It is not the event itself, but it is the person's perception of the event and his or her response to the situation that is the essence of a crisis (Parad 1971). Any person is vulnerable to a crisis at almost any time in their life. Even though others may not see a specific event as disturbing, a crisis reaction is always distressing to the person who perceives the event as a source of intense stress. Situations that cause crisis reactions usually appear suddenly with little or no warning. People are not adequately prepared to manage them. A positive note that can be made here is that crisis reactions are temporary. Most acute crisis reactions subside in 24 to 72 hours.

There are three conditions for a crisis:

1. A highly stressful or hazardous situation.
2. The person senses that the stressful situation will lead to considerable disruption or emotional upset in their normal lives. Thinking shrinks and feelings become dominant (Chapter 4). A state of disequilibrium is established and people in a state of crisis experience *distress, impairment or dysfunction*.
3. The person is unable to resolve the situation by their usual coping methods. The person then feels more out of control (Caplan 1961, 1964, 1969; Roberts 2005).

Most people in a state of emotional crisis have the following five characteristics:

1. They perceive the event as threatening, powerful or critical.
2. They are unable to manage the impact of the event with their usual coping skills.
3. They experience increased fear, tension and mental confusion.
4. They experience considerable subjective discomfort.
5. In a short time, they can proceed to an intense state of crisis or emotional disturbance (Roberts 2000).

An intense state of crisis will appear differently in each person. Relatively common patterns of crisis reactions, however, can be found. People typically react with a combination of the following:

- Helplessness
- Mental confusion and disorganization

- Difficulties in decision making and problem solving
- Intense anxiety, shock, denial and disbelief
- Anger, agitation and rage
- Lowered self esteem
- Fear
- Withdrawal from others
- Emotionally subdued, depressed
- Grief
- Apathy
- Physical reactions such as nausea, shakes, headaches, intestinal disturbance, chest pain or difficulty breathing.

Caution: *Chest pain, difficulty breathing or any other severe physical symptoms should be evaluated by medical staff as quickly as possible* (Mitchell and Resnik 1981, 1986; Salby, Lieb and Tancredi 1975; Slaikeu 1984).

When people reach a state of intense emotional disturbance, their thinking becomes disorganized and unclear. Exaggerated feelings dominate one's reactions. When psychological disequilibrium appears, that is, when a person's thinking ability is suppressed and one's feelings explode out of control, outside *crisis intervention* may be required to rebalance the person and assist them in resolving the situation (Neil et al. 1974; Slaikeu 1984).

Crisis Intervention

Crisis Intervention, as described in Chapter 5, is an *active* but *temporary* and *supportive* entry into the life situation of an individual or group during a period of extreme distress. Its primary goals are to:

1. Mitigate the impact of a critical incident.
2. Facilitate the normal recovery processes of people who are experiencing an emotional crisis.
3. Restore people to an acceptable level of adaptive function.

Crisis intervention is primarily helpful in the most acute phases of a state of emotional turmoil. It is not psychotherapy, however, nor is it a substitute for psychotherapy. Crisis intervention works to alleviate the impact of a crisis experience and to assist people in mobilizing appropriate resources to manage both the situation and the emotional reactions to it. It is encouraging to note that most distressed people react positively to the support provided by others (Stierlin 1909; Salby, Lieb and Tancredi 1975; Mitchell and Resnik 1981, 1986; Everly and Mitchell 1999; Boscarino, Adams and Figley 2005).

Critical Incident Stress Management

Critical Incident Stress Management (CISM) is a package of numerous crisis intervention techniques that are blended and interrelated. CISM is a comprehensive, integrated, systematic and multi-component crisis intervention program. There are features of a CISM program that are in place before a crisis strikes. Pre-crisis education, policy development, training, and planning are among many things that can be done to prepare to manage a crisis experience.

There is also a collection of crisis intervention tools that may be applied during a critical incident. Assessment, strategic planning, large group interventions, individual crisis intervention and advice to management are typical during an event.

Afterwards, crisis intervention techniques include, but are not limited to, small group interventions, individual support, significant other support services, follow-up services and post-incident education. All the parts of a CISM program are interrelated and linked. They are applied systematically which means there is a logical order to the use of interventions (see Chapter 5; Everly and Mitchell 1999; Mitchell 2003).

The most common crisis intervention tools in a CISM program are:

- Pre-incident planning, policy development, education, training
- Crisis assessment
- Strategic planning
- Individual crisis intervention
- Large group interventions (Demobilization, Crisis Management Briefing)
- Small group crisis interventions (Defusing, Critical Incident Stress Debriefing (CISD))
- Pastoral crisis intervention
- Family support services
- Significant other support services
- Follow-up services
- Referral services
- Follow-up meetings
- Post-incident education
- Links to pre-incident planning and preparation for the next crisis.

Critical Incident Stress Management (CISM) Team

There are currently approximately 1,000 Critical Incident Stress Management teams operating in about 30 countries. They often use different names and describe a range of different crisis intervention functions (business staff support, fire service crisis intervention teams, hospital staff support teams, law enforcement teams, crisis response teams for communities, etc.) (Mitchell 2005). Most major airlines and a growing number of air traffic control organizations have CISM or Critical Incident Response Programs (CIRP).

The typical organized crisis intervention team has a mix of members from different parts of one organization or they have representatives from a variety of different organizations who work together. It is not uncommon to have police officers, fire fighters, paramedics, mental health professionals, chaplains, nurses and other CISM-trained support personnel serving on the same team. On an aviation team there may be pilots, flight attendants, ground personnel, managers, flight safety personnel and mental health professionals. Air traffic control organizations generally have their own teams, but they are more than willing to combine their efforts with other teams if the circumstances warrant their involvement. When a critical incident of sufficient magnitude occurs, it is a common practice for many crisis response teams and organizations to unite into a task force to provide the best assortment of crisis intervention services.

On most teams, which have combined many organizations into a unified system, when they receive a call for assistance, specific personnel are deployed from the team. For example, if law enforcement support services are required, only law enforcement personnel and specified mental health professionals are deployed to manage the situation. Likewise, in aviation, if a flight deck crew is dealing with a traumatic incident, only pilots and specified mental health professionals would be deployed from a CIRP team. Flight attendants would be served by other flight attendants and air traffic control personnel would help air traffic controllers experiencing the aftermath of a traumatic event. There can be situations in which a crossover is necessary or helpful (for example a mixed team of pilots and flight attendants serving a distressed crew of an airliner). Use whatever combination of team members and tactics that appears to be helpful to support a distressed crew. These combinations are part of a strategic approach that matches resources to the needs of a group and to the individuals within that group.

There is at least one major advantage in having all the personnel in a specific geographical area belong to the same team. In the event of a large scale event peers from different professional groups who belong to the same team can work together. Other advantages include joint training, sharing of resources, joint practice sessions and case reviews for continuing education (Mitchell 2003, 2004; Mitchell and Everly 2001).

The Primary Principles of Crisis Intervention

Members of a CISM or CIRP team must adhere closely to a basic set of operational principles when providing crisis intervention services of any type. The paragraphs below describe the essential principles in all crisis intervention.

Seven primary principles have emerged during the 100-year history of crisis intervention. *Proximity* is the first principle of crisis intervention. Crisis work is often provided in surroundings familiar to the victims. In the aviation world, many crisis intervention tactics are applied at airports or in some cases on board a parked aircraft. The principle of proximity emerged from the battlefield experiences of

World War I. When soldiers were given crisis intervention services close the front lines 65 per cent returned to combat in three to four days. If, instead, they were given the same support services in a rear area far from the battlefield only 40 per cent of them returned to combat and it took three to four weeks to achieve that goal (Salmon 1919).

The second principle is *immediacy*. People in a crisis cannot wait long for help. The longer they wait, the less likely crisis intervention will be effective (Lindy 1985). The third principle is that of *expectancy*. Early in the intervention, the provider of crisis intervention has to instill hope that it is possible to manage and resolve the situation (Salmon 1919; Kardiner and Spiegel 1947; Solomon and Benbenishty 1986). The fourth principle is *brevity*. No one has the luxury of abundant time in a crisis so crisis management actions have to be brief (Parad and Parad 1968).

The fifth principle is *simplicity*. People do not handle complexity very well in the midst of a crisis. Therefore, crisis interventions have to focus on the ease of applying the solutions developed during the crisis intervention contacts. Simple, well thought out interventions will be the most effective in the majority of cases.

Crisis workers must have creativity when they are working in extraordinary and challenging circumstances. The ability to be *innovative* in the face of unusual, threatening, and disturbing situations is a key to good crisis intervention and the sixth major principle of crisis intervention. Appropriate helpful interventions may have to be developed on the spot. Finally, the seventh principle is that whatever actions are chosen in a crisis should be *practical*. Impractical solutions are no solutions at all. People struggling through a painful experience will see them as insensitive and uncaring.

The Seven Core Principles of Crisis Intervention:

1. Proximity
2. Immediacy
3. Expectancy
4. Brevity
5. Simplicity
6. Innovative
7. Practical

(Lindemann 1944; Kardiner and Spiegel 1947; Artiss 1963; Caplan 1961, 1964, 1969; Parad 1971; Neil et al. 1974; Duffy 1979; Slaikeu 1984; Lindy 1985; Everly 1999; Roberts 2000, 2005).

Several other principles can serve you or your crisis response team well when you are helping someone in the aviation industry who is enduring a crisis. The first is *stay within your training levels*. Do not attempt to do things for which you have not been adequately trained. If you lack the experience necessary, call for help.

We advise the crisis intervener to *avoid 'unboxing' anything that cannot be 'reboxed' in the time available*. In other words, do not engage in conversations about material that you do not have the time to listen to in its entirety. It can be quite disturbing to a troubled person for a helper to try to engage in a discussion of complex and painful personal information only to cut off the distressed person part way through the discussion because there is insufficient time. It is better to say something like this:

> This sounds like we really need to sit down and have thorough discussion of your situation. I want to be helpful so it is best if we can meet later today and spend some time on this problem. That way we can thoroughly discuss it and help you to work out a solution. My schedule is free between 1:00 p.m. and 5:00 p.m. Can we meet today at 1:00 p.m.? I will have much more time available then.

Although it is often therapeutic, crisis intervention is neither psychotherapy nor a substitute for psychotherapy. That point is so important that it is repeated several times in this chapter. Crisis intervention cannot handle deep-seated psychological problems. One cannot achieve substantial, long-lasting life changes in the midst of a crisis. Keep your expectations reasonable. *Do not encourage discussions of excessive details of old psychological material when dealing with a crisis.* Focus on the *here and now*. Delving deeply into excessive details about a past situation is time consuming, unhelpful and often counter-productive. If you encounter someone who has complex psychological issues, it is better to make a referral to a mental health professional (Everly and Mitchell 1997, 1998).

After crisis intervention a reasonable degree of improvement or movement toward recovery is usually apparent and, if it is not, consider a referral. Crisis intervention is by nature brief. We should see some calming of the person and the development of a crisis plan as well as some efforts to resolve the problem in a fairly short period of time. Positive results typically appear with a minimal number of contacts (Boscarino et al. 2005). Typically, crisis intervention involves between three and five contacts. Some individual crisis intervention contacts last 5 to 15 minutes.

Occasionally an individual contact may last about 45 minutes. Referral is seriously considered when the number of contacts progress beyond a total of five. If the situation does not resolve by eight contacts, then a referral is clearly necessary. Further delay is unhelpful and may even be dangerous to the person's well-being. Crisis intervention should only last long enough to be helpful. Make a referral immediately if the distress is extreme, if no calming or recovery is evident, or if serious impairment persists beyond the first three contacts.

General Guide for Crisis Intervention Contacts:

- *Three to five brief contacts are typical* (for example any combination of individual crisis intervention, small group support services and individual follow-up contacts including telephone support or brief visits).
- *Six to seven contacts may be necessary in some cases.* If no progress has been made by the fifth contact, a referral for professional care becomes more likely. Referral for professional care should be seriously considered.
- *Eight or more crisis intervention contacts* – professional referral is indicated.

Keep the primary objectives of crisis intervention in mind. Crisis intervention is a limited helpful strategy. Do not establish outcome expectations that are far beyond its capabilities. Crisis intervention cannot cure disease or significant mental disturbance. It is not a cure for Post Traumatic Stress Disorder (see PTSD sections below). The primary objectives for crisis intervention are:

1. Mitigation of the impact of a critical incident or traumatic event
2. Facilitation of recovery processes
3. Restoration to adaptive function
4. Identification of individuals who may need additional support or a referral for psychotherapy
5. Support and secondary prevention (Mitchell and Resnik 1981, 1986; Mitchell and Everly 2001).

Critical Incident Stress Management (CISM) is primarily a subset of the field of crisis intervention. A subset is defined as a collection of elements within a certain category that are clearly related to each other and all of which can be found within a larger *umbrella* category. All of the elements contained within a subset (CISM) can be found within the main category (crisis intervention). From its inception, Critical Incident Stress Management (CISM) was deeply rooted in the field of crisis intervention.

Therefore, the many elements of a CISM program are all crisis intervention elements and they share the same history, theoretical foundations, principles, goals, strategies, procedures, methods and techniques associated with crisis intervention (Mitchell 2004).

Since CISM is a subset of crisis intervention, the reader can accurately assume that every time the term, *CISM* is used in this book there is an implication that crisis intervention principles and practices are being discussed. Please note that current literature refers to crisis intervention as *early intervention*, *psychological first aid*, and *emotional first aid* (American Psychiatric Association 1964; Neil et al. 1974; Mitchell 2004).

Crisis Intervention is *Not* Psychotherapy

Crisis Intervention is psychological *first aid* and focuses on support, not cure. It is not psychotherapy nor is it a substitute for psychotherapy (American Psychiatric Association 1964; Neil et al. 1974; Everly 1999, Everly and Mitchell 1997, 1999; Mitchell 2003).

CISM or CIRP have many crisis intervention tools that can be applied by skillful team members to assist people in a crisis. The most common form of crisis intervention is individual emotional *First Aid*. Some describe it as a *one-on-one* support service. It is presented here to help emphasize the fact that crisis intervention is not psychotherapy (see Table 3.1).

Table 3.1 Psychotherapy and crisis intervention

Differences between Psychotherapy and Crisis Intervention	
Psychotherapy	**Crisis Intervention**
Context	
1. Repair of existing psychological damage	1. Prevention of psychological damage
2. Reconstruction of self-esteem	2. Acute stress mitigation
3. Personal growth	3. Restoration to adaptive function
Strategic foci	
1. Conscious and unconscious sources of psychopathology	1. Conscious processes
	2. Environmental stressors
	3. Situational factors
Location	
1. Safe, secure environment	1. Close proximity to stressor
2. Therapy office	2. Homes, businesses
3. Hospital facilities	3 Anywhere needed
Purpose	
1. Personal growth	1. Emotional *first aid* to
2. Change of life experience	2. Mitigate the impact of an event
3. Alteration of behavior	3. Assist the person in crisis to return to a state of adaptive functioning
4. Reduce family dysfunction	4. Enhance unit performance and unit cohesion when working with groups
Temporal focus	
1. Past	1. Immediate present
2. Some Present	2. *Here and now*
3. Some future	
Providers	
1. Mental health professionals	1. A trained, outgoing person who cares for people and has a desire to help those in a state of crisis
	2. Paraprofessionals (people who work in parallel with mental health professionals) such as aviation industry employees, emergency medical personnel, fire fighters, police officers, doctors, nurses, soldiers, emergency communications personnel, etc.
	3. Clergy
	4. Mental health professionals
Provider Role	
1. Guide	1. Protector
2. Collaborator	2. Active listener

3. Consultant	3. Offering directions
4. Observer	4. Facilitator
5. Commentator	5. Educator
Timing	
Typically within weeks to months or years after the development of a problem that interferes with normal life pursuits; delayed, distant from stressor	During a critical incident and in the immediate aftermath of an exposure to the event; immediate, close temporal relationship to stressor
Duration	
8 to 12 45-minute sessions for short-term months to years of weekly sessions for as long as needed for long term	3 to 5 contacts some of which are only minutes in length maximum contacts: usually 8
Goals	
1. Symptom reduction	1. Stabilization
2. Reduction of impairment	2. Mitigation of impact
3. Correction of pathological states	3. Reduction of impairment
4. Personal growth	4. Mobilization of resources
5. Reconstruction	5. Normalization of experience
	6. Restoration of adaptive function
	7. Resolution of immediate crisis state
	8. Referral to next level of care

A Guideline for Individual Psychological *First Aid*

Dr. George Everly developed an approach to individual crisis intervention and he called it the SAFER-R model (see Chapter 5). Each letter in the word, SAFER-R has a meaning.

S stands for stabilize. One of the first objectives when helping a person in a crisis is to *stabilize* the situation by eliminating the stimuli that might make the situation worse.

The support person next *acknowledges (A)* the disturbing event and lets the person know that the helper is aware that the person is experiencing significant distress.

The *F* stands for facilitate. The helper asks the distressed person to briefly describe what happened and their personal reactions to the event. The helper then needs to *facilitate* the exploration of options and the development of a plan of action.

The CISM team member then *encourages (E)* and assists the person in the crisis to put the crisis plan into action. This step in the process also offers an opportunity for the team member to teach some stress management information and provide some explanation of what is happening now or is likely to happen as the distressed person progresses through the crisis.

The helper stays involved until *resolution (R)* of the crisis is evident or until an appropriate *referral (R)* for additional support or professional involvement is made (Everly 1999; Everly and Mitchell 1997, 1998).

The SAFER-R Model of Individual Crisis Intervention – A Summary:

S   *Stabilize*, provide privacy, cut stimuli
A   *Acknowledge*, a bad event has occurred and the person is in crisis
F   *Facilitate* the conversation and the exploration of the options and a plan
E   *Encourage* the person to take action to help themselves
R   *Resolution* is evident in reduction of distress and actions taken
R   *Referral* is used if reaction is extreme or if no resolution is achieved.

Aviation Critical Incidents and Critical Incident Stress

Aviation employees experience the many critical incidents that virtually every adult experiences in the course of their life and work. There are, however, critical incidents that are specific to the aviation industry. These critical incidents challenge, and frequently overwhelm, the coping abilities of individual employees or entire aircraft crews. They are (Mitchell 1982):

- Work related deaths of colleagues
- Serious work related injuries to colleague
- Suicide of a colleague
- Disasters involving aircraft
- Accidental killing or wounding an aviation employee or a passenger
- Events involving a high degree of personnel threat
- Aircraft *close calls* in the sky or on the ground
- Severe turbulence
- Seriously ill or injured child on board
- Major medical emergency while airborne
- Threatening passenger
- Violence from any source while in flight
- Aircraft damaged or impaired while in flight
- Fire on board an aircraft
- Hazardous materials leak on an aircraft
- Bomb threat
- Other distressing experiences.

Critical Incident Stress

Whether they are the powerful general events that impact most people or the aviation specific situations listed in the previous section, critical incidents are the beginning point of the critical incident stress reaction. As noted earlier, critical incidents cause an acute state of emotional turmoil called a *crisis*. *Critical Incident Stress* is the cognitive, physical, emotional, spiritual state of arousal that accompanies the crisis response. When thinking, feelings, bodily functions and belief schemes become aroused by exposure to a critical incident, behavior changes. For example, if a person becomes frightened, he or she may run away or hide. Running and hiding are behaviors that occur as a result of the crisis reaction.

It is reasonable to expect that when the event that caused the crisis reaction ends, the elevated state of arousal should subside. That is exactly what happens in most crises. The crisis reaction usually calms as the situation settles down. In some cases, however, the stimulus that produced the crisis reaction is either extremely powerful or it generated such intense reactions that they linger long after the critical incident has faded into history. Critical incident stress reactions that are generated by an intense traumatic event often continue from several hours to days and even several weeks after the event ends. As response to a severe traumatic experience, critical incident stress does not easily shut off. Memories of the traumatic situation are intense and disturbing and they continue to put mental and physical systems on *alert*. They trigger cognitive, physical, emotional and spiritual arousal as if the event had not ended (Mitchell and Everly 2001; Mitchell 2004).

Here are some common signs and symptoms of critical incident stress:

1. Common Stress Signals
   - Preparations for fleeing or fighting
   - Increased arousal and readiness
   - Automatic activation or suppression of body functions
   - The mind reviews the past for experiences that help now
   - More rapid breathing rate
   - Yawning (compensates for lower oxygen levels in the body during times of stress)
   - Increase of natural stress chemicals in the body
   - Body senses are more alert to danger
   - Eye pupils open wide and the eyes scan for danger
   - Seeking defensive positions
   - Heart rate increases
   - More active listening
   - Increased anger and irritability
   - Aggressive stance
   - Seeking cues to danger or safety.
2. Arousal Signals: Moderate Stress Reactions
   - Emotions become more hair trigger
   - Intense anger and irritability
   - Frustration
   - Lowered attention span and forgetfulness
   - Difficulties concentrating
   - Problem solving difficulties

- Emotional depression
- Risk taking behaviors
- Increased use of alcohol
- Withdrawal from others.

3. Arousal Signals: Excessive Stress
   - Destructive changes in personal relationships
   - Paranoid feelings (someone is trying to hurt you)
   - Intense depression (chronic feelings of sadness)
   - Negative changes in physical and mental health
   - Changes in personality
   - Deterioration in performance
   - Feelings of intense pressure
   - Chronic irritability
   - Loss of objectivity
   - Extreme frustration
   - Thoughts of suicide.

4. Arousal Signals: Extreme Stress
   - Loss of control
   - Excessive use of alcohol and other drugs
   - Severe depression
   - Rage reactions (anger out of control)
   - Homicidal feelings
   - Striking out at others
   - Physical breakdowns
   - Mental breakdowns
   - Suicidal actions.

5. Dangerous Signals of Stress.

These signals require immediate, corrective action since health and safety is in danger.

   a. Physical
      - Chest pain or difficulty breathing
      - Excessive blood pressure
      - Excessive dehydration
      - Collapse from exhaustion
      - Unusual heart beat patterns
      - Signs of severe shock (weak pulse, pale color, excessive sweating, and rapid breathing)
      - Excessive dizziness
      - Heat stroke, vomiting or blood in the feces.
   b. Mental
      - Mental confusion and a decreased awareness of surroundings
      - Disorientation to place, time and person

- Difficulty concentrating and making decisions
- Poor problem solving when under pressure
- Hyper-alert (excessive seeking out of threats)
- Racing thoughts and problems focusing on one thought at a time
- Disrupted thinking
- Poor recall of names and functions of familiar objects.

c. Emotional
- Severe fear and panic *attacks*
- Numb, shock like state
- Loss of emotional control and exaggerated emotions
- Uncontrolled rage
- Overwhelming sadness.

d. Spiritual
- Hopelessness
- Despair
- Shattered self-esteem or feelings of self-efficacy
- Religious obsessions
- Religious compulsions
- Religious hallucinations
- Religious delusions.

e. Behaviors
- Slurred, unintelligible extremely slow or rapid speech
- Crying spells
- Uncontrolled laughter in the face of serious threats
- Excessive anger at minor issues and violent outbursts
- Destruction of unit's property
- Curling up, hiding, startle response, or continuously rocking motions
- Excessive pacing, wringing of hands, body or facial tremors
- Suicidal actions.

This is a long list, but it is by no means a complete list. Any cognitive, physical, emotional or spiritual symptoms can be caused by or made worse by a traumatic experience. Critical incident stress begins as a normal response of normal people to an abnormally intense event. In most cases, it will reduce over time and by about three weeks the reaction is usually concluded. Memories of the traumatic event, however, may remain for a lifetime. If critical incident stress remains unresolved for longer than a month, a more serious condition called *Post Traumatic Stress Disorder* may result (Everly 1999; Mitchell and Everly 2001; Flannery 2004; Mitchell 2005).

Post Traumatic Stress Disorder

This chapter focuses on managing critical incident stress in aviation industry employees. Every effort should be made to mitigate and resolve critical incident stress as quickly as possible. It is far easier to manage critical incident stress reactions and get them resolved than to have to treat something worse such as Post Traumatic Stress Disorder (PTSD; see Chapter 4). Unfortunately, some people in the aviation industry develop PTSD when their critical incident stress reactions are not resolved. A basic knowledge of PTSD can be enormously helpful when such cases emerge in aviation.

Remember the starting point for PTSD is a *critical incident*. PTSD is *not* an automatic result of exposure to a horrible, awful, terrible, grotesque or terrifying critical incident. Some critical incidents may trigger strong stress reactions, but, these stress reactions are normal and adaptive. They occur in the healthiest of people. The critical incident stress response should, in fact, turn *on* when there is an actual or perceived threat. That is the expected reaction. When the threat subsides, the critical incident stress reaction should then turn *off*.

PTSD is an extreme stress reaction to an exceptionally stressful critical incident and it develops only when the normal critical incident stress reaction is not resolved. PTSD is, in essence, a *super strength version* of an acute stress response that became stuck in the *on* position.

Estimates are that nearly 90 per cent of adults have at least one intense traumatic event in their lifetime. Yet, not everyone develops PTSD. The diagnosis emerges in about 5 per cent of men and about 10 per cent of women in the general population. Exposure to certain severe stressors is associated with higher rates of PTSD. In the specific populations listed below, for example, the rates are higher, depending on the type and intensity of the trauma.

- Disaster victims   34%
- Sexual assault   49%
- Captivity, kidnapping   53%
- Urban fire fighters   15 to 31%
- Urban police officer   13 to 30%
- Viet Nam veterans   15 to 30%
- Afghanistan and Iraq war veterans   12 to 17%
- Torture victims   50 to 90%

There are many reasons why some people develop PTSD and others do not. Typically the individual person's mental interpretation of the event is a more important factor in the development of PTSD than the event itself. Strong feelings of personal responsibility for what happened or intense feelings of guilt are among the most frequent contributing factors. It should be noted that intense guilt feelings can complicate the recovery process.

A combination of other factors also influences the development of PTSD. A greater number of exposures to past horrific events or disturbing aviation related situations combined with a greater degree of personal involvement in a current traumatic event can certainly set the stage for PTSD to develop. Concurrent life stressors such as pregnancy, illness, death of a loved one, or other major stress may have an influence on whether a person develops PTSD or not.

No one should discount the role of an individual's personality in PTSD. Highly anxious people are more prone to develop PTSD. A history of severe traumatic events especially overwhelming childhood traumas enhance the potential for PTSD. It is also known that feeling as if one was *outside* his or her body or *in a movie* during the traumatic experience has been associated with higher potential for PTSD (Everly and Lating 2004; Antonellis and Mitchell 2005; Flannery 2004; Mitchell 2005).

Criteria for PTSD

There are six criteria for PTSD. All six must be present for a mental health professional to make the diagnosis. As you review these criteria, which are presented here for informational purposes only, you should keep in mind that every person reacts to traumatic events in their own way. The criteria are only guidelines and self diagnosis is not suggested. A person who suspects that he or she is suffering from PTSD would be wise to consult a mental health professional to assure proper assessment and treatment, if necessary.

Members of a CIRP or CISM team within the aviation industry may use these criteria to help develop a decision to refer a person to a mental health professional for further evaluation. The criteria for PTSD are:

1. *Exposure* to a horrible event. Usually there is a threat of death or serious injury in the situation and intense fear, helplessness and horror result. Some disturbing aviation incidents such as crashes, threatening and violent passengers and severe, injury-generating turbulence fit this criterion. Exposure may be direct or witnessed.
2. Intense *arousal symptoms* (restless, sleepless, hyper-alert, unable to relax, difficulties concentrating). Arousal symptoms suggest excessive physiological activation and heightened psychological vigilance.
3. *Intrusive symptoms* (mentally see, hear, feel, smell, taste aspects of the event and have repeated bad dreams or nightmares). Intrusive symptoms are simply a mental replay of the traumatic experience. The human brain categorizes traumatic stress memories as intense, immediate and permanent. It takes time to calm the brain system once a traumatic events ends. The intense memories, therefore, remain quite active for some time after the traumatic event.
4. *Avoidance symptoms* (shutting off one's emotions, avoiding places, equipment, people, conversations and other stimuli that remind us of the event).
5. PTSD *symptoms last beyond 30 days*. Anything short of a month is considered Critical Incident Stress (the normal, but distressing, response of normal people to an abnormal event). After a month PTSD may be present. Only an appropriate evaluation by a competent mental health professional can confirm

or rule out the diagnosis of PTSD.
6. The PTSD condition causes *significant disruption to and impairment of normal life pursuits*. Social life, work and home life may be impaired (Everly and Lating 2004; Flannery 2004; Antonellis and Mitchell 2005).

Variations in PTSD Severity

When PTSD does develop, it does not always affect every person in the same way. People are different from one another. Some develop mild variants of PTSD. Other people may have moderate or severe variations of the condition. If a person's PTSD is relatively mild to moderate, some uncomfortable symptoms may exist. People tend, however, to remain functional and reasonably productive. In fact, with the passage of time, mild to moderate variations of PTSD may spontaneously resolve with little or no professional assistance. When PTSD is severe, it is very difficult to manage on one's own. It can be life altering in a very negative sense. People with severe PTSD have trouble performing daily life functions. They find it difficult to work, care for a family and have the normal relationships of everyday life. They withdraw from others and often use alcohol or drugs to numb their pain. Very often professional guidance is required to manage a severe level of PTSD (Everly and Lating 2004; Flannery 2004; Antonellis and Mitchell 2005; Mitchell, 2005).

Complications in Managing PTSD

Human beings do not live in a world in which only one terrible thing happens at a time. PTSD is particularly challenging to recover from if the disorder combines with any of the following:

- Multiple traumatic events over time
- Work related traumatic events combined with personal traumas
- Alcohol and drug abuse
- Grief
- Serious emotional depression
- Disabling physical injuries
- Memory and thinking problems
- Other mental disorders
- Physical illness.

Treating PTSD

PTSD is certainly treatable. Millions of people have worked their way through the condition and are functioning well in their work and personal lives. This author is aware of numerous cases in which aviation personnel received successful treatment and are flying or working in other areas of the industry today. There are several conditions, however, that are crucial for successful PTSD treatment:

1. Positive motivation and drive from the person with PTSD.
2. A trained and qualified mental health professional.
3. The right treatment program which is carefully constructed to match the needs of the person with PTSD.
4. In the aviation industry, particularly in regard to pilots (Chapter 8) and air traffic controllers (ATCOs, Chapter 7), the therapist must be familiar with the nature of the job.

There are several successful PTSD treatment programs. A few are described in the paragraphs below. The most effective treatments are available only from people with at least a master's level degree in counseling or clinical psychology, social work or psychiatry. The most helpful therapists are those who are thoroughly familiar with PTSD and experienced in its treatment. Aviation personnel are encouraged to learn as much as they can about specific therapies before entering a therapy process. Careful assessment by a knowledgeable mental health professional is a crucial step in the therapy process. It should take place as soon as possible after a person recognizes that their stress reactions are remaining intense beyond a reasonable time (Everly and Lating 2004; Flannery 2004; Antonellis and Mitchell 2005).

1. Behavior Therapy (aims at recognizing stress reactions in the body and controlling those reactions):
    a. Relaxation training (has been well received by aviation personnel who report that being able to apply some learned stress management techniques helps a person to feel less out of control and more hopeful of resolution)
    b. Biofeedback
    c. Systematic desensitization
    d. Many reports of symptom reduction exist for this kind of therapy.
2. Cognitive Behavioral Therapy. Research indicates CBT is among the most effective therapies:
    a. Helps a person understand the meaning behind the trauma
    b. Challenges errors in facts and beliefs
    c. Pairs new thoughts and associations with trauma
    d. Event feels less upsetting and symptoms decrease.
3. Pharmacotherapy:
    a. Under certain circumstances, medications may be administered to people who are overwhelmed with symptoms of distress or who have several

simultaneous psychological conditions
b. The right doses of the right medications are required
c. Medication therapy can be expensive
d. Medications rarely cure PTSD, but medications do help to control painful symptoms
e. Drug dependency is a possibility when some drugs are prescribed
f. Side effects from certain medications are also possible
g. Compliance with the treatment protocols is mandatory.

Therapies Developed Specifically for Trauma

Several beneficial trauma therapies were developed in the 1980s and 1990s. These therapies aim specifically at traumatic stress and a growing body of positive outcome research supports their application.

1. Eye Movement Desensitization and Reprocessing (EMDR):
   a. Developed specifically for trauma treatment
   b. Both recent and distant past traumatic experiences have been successfully treated with EMDR
   c. Therapist guides person through a series of eye movements combined with verbal discussions
   d. Biological and intellectual processing of the traumatic event apparently occurs
   e. Person integrates the traumatic experience into their memories
   f. Memories of the traumatic event usually remain, but the intensity of the emotions generated by the memories is notably reduced
   g. Relief often occurs in a few sessions
   h. Research is very strong for positive effects.
2. Trauma Incident Reduction (TIR):
   a. The therapist has the person tell the trauma story repeatedly
   b. Repetition of the traumatic event seems to lessen the emotional intensity of the memories
   c. May take several sessions to achieve results
   d. Numerous positive reports of success.
3. Thought Field Therapy (TFT):
   a. Limited research so far, but more research is underway
   b. Some controversy exists over its effectiveness
   c. Some therapists and clients report success in the form of symptom reduction

d. The therapy uses a series of self-taps or rubs on the face, hands, chest while thinking of the traumatic event (Antonellis and Mitchell 2005; Mitchell 2005).

A good therapist has a number of tools in the therapeutic *toolbox*. Some therapies work better on some people than on others. If one therapy fails to work, the therapist must move onto another until something or some combination of interventions is successful. The benefits of any therapy should show up in about three months (Everly and Lating 2004; Flannery 2004; Antonellis and Mitchell 2005; Mitchell 2005).

Strategic CISM Programs for Aviation

Strategy is the art and science of maneuvering one's resources into the most advantageous position in order to improve the chances of a successful outcome. It implies that the right people provide the right services to the appropriate targeted populations in the best possible time frame and under the most advantageous circumstances. An appropriate strategy is one reason why the Critical Incident Response Program (CIRP) or Critical Incident Stress Management (CISM) programs have been so well received by the aviation industry. They contain all of the elements of a sensible strategic approach to crisis intervention services. Here are the key elements of a sensible strategic psychological first aid program for aviation:

- *Assessment skills*. The CIRP or CISM team should be able to asses both the severity of the incident and the level of distress in the personnel who were involved in or who witnessed the traumatic event.
- *Strategic planning skills*. The CIRP team members should be able to develop a comprehensive, integrated, systematic and multi-component (CISM) approach to assisting traumatized individuals or groups within aviation.
- *Multi-tactic approach*. No one crisis intervention service will be suitable to all people at all times and under all circumstances. A multi-component approach is the only realistic and sensible approach that is likely to be effective in crisis intervention within the aviation industry (see the sections on Critical Incident Stress Management earlier in this chapter).
- *Follow-up services*. Once a series of crisis intervention services have been provided, it is essential that CIRP or CISM team members reconnect with individuals or group members to assure that they are in fact recovering from the traumatic experience. Telephone calls, visits to the work site or to a person's home are crucial follow-up crisis intervention techniques in the field of emotional or psychological first aid.
- *Linkages to professional care*. The vast majority of people in the aviation industry do not need professional psychological care. It would be irresponsible, however, not to have in place links and referral mechanisms to professionals for the few who would benefit from these services. In addition, CIRP or CISM

team members need professional oversight and guidance to assure that their psychological first aid services are the very best available.

The Strategic Formula

There are five issues that must be addressed to make the best strategic plans for crisis intervention services. They are:

1. *Target(s)*. Who is in need of crisis support services (individual, crew, passengers, ground personnel, ticket agents, other)?
2. *Type(s)*. What interventions are the most appropriate? Remember, CISM is a blend of services. Even if one type of intervention is selected for an individual, it is still necessary to follow-up with the person and make sure that the person is recovering.
3. *Timing*. Some psychological first aid or crisis intervention services can be provided immediately after the traumatic event. Others are best applied a day or two later. It is essential that whatever crisis services are applied be carefully timed to have the most beneficial effects.
4. *Themes*. Themes are any circumstances, situations, issues, concerns or considerations that influence decision making, timing or the selection of crisis interventions. Themes involve issues like schedules, relationships between crew members, company policies, circumstances of the traumatic event, death or injuries to crew personnel and a host of other considerations. Intelligence gathering on these issues is crucial if the CIRP or CISM team wants to choose the right interventions to match the needs of the individuals or group members to be served.
5. *Team*. When target, type, timing, and themes have been considered, the CIRP or CISM team must decide which team members should be deployed to provide the services. The right choices of interventions at the right time will not be helpful to people in a crisis unless the team members deployed to provide services are the best match (Mitchell and Everly 2001).

The response levels described below should be helpful in demonstrating how the five issues in strategic planning are tied together. An aviation industry response to a traumatic event will typically fall under one of six levels.

Response Levels

Level One A – Aviation Personnel – Small Scale Event

Theme: Aviation related incident (significant air space incursion, severe turbulence, injury to another crew member, violent passenger threatening a crew member, etc.).

Target: Aviation personnel such as flight crew, flight attendant crew, air traffic controllers, ground personnel. One to four people requiring assistance.

Type: Individual crisis intervention (psychological first aid) and/or a brief crisis oriented conversation with those needing assistance (see SAFER-R). Assess the situation, mitigate the disturbing factors, mobilize resources, normalize the person's reactions, and restore adaptive functions.

Timing: Hours to days after the event, depending on the needs of the aviation person or personnel involved and the circumstances.

Team: (refers to the number and specialty area of team members deployed). Pilots should be assigned to assist pilots. Flight attendants help other flight attendants, air traffic controllers are assisted by air traffic controllers and so on (within reason). One or, preferably, two CIRP or CISM responders according to their specialty areas or functions.

Level One P – Aviation Personnel Involved in a Personal Crisis

Theme: Individual employee experiences a non-work related crisis (family emergency, death in the family, sick family member, severe personal illness, direct threat of violence, loss of job, etc.).

Target: Employee impacted by an external loss, threat or other crisis.

Type: Individual crisis intervention, mobilization of resources, support.

Timing: As immediate as possible after the crisis situation occurs.

Team: One CISM/CIRP team member who is reasonably close to the place where the person in crisis is located.

Level One C – Customers Need Assistance from Airline Personnel after a Significantly Distressing Event

Note: Even crisis teams that claim to be in place only for services to their own colleagues, occasionally end up assisting members of the public when certain circumstances arise. It is best to be prepared for the unusual request for assistance.

Theme: Passengers are distressed by an aviation-related disturbing event (severe turbulence, hard landing, threatening mentally disturbed person on board).

Target: Passengers, non-aviation groups, heterogeneous groups.

Type: Crisis Management Briefing (CMB).

Timing: As close to the ending of the event as is reasonably possible. Typically within one hour but may be done a little later if circumstances do not allow it earlier.

Team: one CIRP or CISM team member and one representative of the airline. Objective is to provide general information, practical guidelines and stress management information.

Level Two – Significant Small Scale Event Involving both Flight Deck Crew and Flight Attendant Crew or an Entire Air Traffic Control Center's Shift

Theme: Heart attack (surviving) or serious injury of a crew member or air traffic controller during the performance of duties. *Close calls*, severe turbulence, damage to aircraft while airborne, fire on board or ground accident causing injuries are also examples.

Target: Flight deck and cabin crews, air traffic control shift, ground personnel, etc.

Type: Defusing, followed by individual crisis intervention, and follow-up with a CISD if necessary. Confirm resolution of crisis by means of individual contacts. Refer if resolution not achieved.

Timing: Start immediately (within reason) after situation ends.

Team: A team of two for the defusing; three for the CISD including a mental health professional trained in CISM. Follow up contacts are on an individual basis.

Level Three – Death in the Workplace

Any event causing the death of an aviation employee should always be managed as a serious event needing considerable CIRP or CISM team involvement.

Theme: Employee(s) die in the performance of their duties.

Target: May have to be designated as separate groups depending on numbers of employees who were involved and the circumstances of their involvement. Smaller groups (about 30 or below) are ordinarily managed together. Larger groups (about 40 or more) may need to be split into more manageable groups according to their level of involvement during the incident.

Type: Day one five phase CISD; individual crisis intervention; family support services; Crisis Management Briefings to larger less involved groups; individual assistance during the funeral; Seven phase CISD after the funeral; more individual crisis intervention services; follow up meetings.

Timing: Support services begin immediately after the event and may continue for several weeks. Teams must be alert to anniversary times which generate additional emotional reactions.

Team: The entire team is likely to be involved in several different interventions over the weeks following the death. If the team members were too emotionally close to the deceased person supplementary crisis team members need to be borrowed from other teams.

Level Four – Major Incident (Large Scale) – No Aviation Personnel Deaths, but Serious Injury and Possibly Deaths to Customers and/or Bystanders; Major Destruction to Aircraft or other Equipment

Theme: Terminal fire; aircraft off the end of the runway; bomb in vicinity of air traffic control center; fire on board aircraft with passenger deaths; violent passenger who harms others; incident with high media involvement.

Target: Multiple targets – crew, air traffic control personnel, passengers; family members waiting for arrival of loved ones, etc. Groups are separated from one another and prioritized according to need. Services are chosen according to appropriateness for the specific target populations.

Type: Crisis Management Briefings; Defusing; individual crisis intervention; CISD; family support services; follow-up services; post incident education, etc.

Timing: Interventions begin as soon as reasonably possible once the event ends.

Team: The entire team is most likely to be mobilized to manage an event of this magnitude. In some cases additional support may be requested from other aviation based teams.

Level Five – Air Crash Disaster. Multi-fatality/Multi-casualty Event. Large Scale Event

Theme: Major air crash; multi-casualty event; deaths / injuries to multiple crew members; deaths/injuries to multiple passengers or people on the ground.

Target: Multiple target populations (handle separately); surviving crew; surviving passengers; air crash rescue and fire fighting personnel; ground personnel; colleagues; family members; family assistance personnel from the airline; witnesses; community members, etc.

Type: Crisis Management Briefings; demobilizations; defusings; individual crisis intervention services; CISDs; follow-up services; family support services; follow-up meetings; referrals, etc. All CIRP or CISM services should be applied according to need and as appropriate.

Timing: Interventions should be begun as soon as reasonably possible after the event. CISDs are much later. Many other services are needed earlier. Services may continue for weeks to months after a major air disaster.

Team: All local CIRP or CISM team resources will be quickly overwhelmed. Assistance from numerous other airline teams or air traffic control centers will most likely be necessary. In some cases teams from other agencies and organizations will be necessary.

Level Six – A Catastrophic Event (Multiple Losses of Aircraft. Air Crash into a Densely Populated Area. Wide Scale War.)

Level Six – Catastrophic event similar to the events of 11 September 2001. The catastrophic event would more than likely generate a multi-agency, multi-airline, multi-jurisdictional CISM/CIRP response. Every available team member from multiple teams from many cooperating organizations will need virtually every crisis intervention tool and for a considerable period of time. Several countries may have to share their CISM or CIRP resources. See the Theme, Target, Type, Timing, Team profile for Level five.

Conclusion

A considerable body of current literature suggests that crisis intervention, in the form of a Critical Incident Stress Management program, has positive effects on the reduction of stress symptoms. Such programs may play a preventative role against the development of long-range psychological problems (Bohl 1991, 1995; Jenkins 1996; Wee et al. 1999; Chemtob et al. 1997; Boscarino 2005; Deahl et al. 2000; Richards 2001; Campfield and Hills 2001, Vogt et al. 2004).

Each study provides a window of light into the field of crisis intervention. Each study helps us to understand what enhances the potential for successful and helpful interventions and what may decrease the potential of success. Even without formal studies the people who have received crisis intervention support tell us that the assistance is welcome and effective.

A personal note may help here to emphasize this point and round out this chapter. Very recently I was flying home from an assignment. I nodded off to sleep during the flight and awoke to see a flight attendant standing over me and reading the CISM team name on my shirt. She smiled and apologized as if she had awakened me (which she had not) and she stated that she had had to use the services of her airline's CISM team twice within a few weeks. She told me that they were wonderful people who had helped her enormously. One incident involved a passenger death on board one of her flights. The other was the resuscitation of a small baby while in flight. She stated that the team members who helped her with individual CISM services had saved her career and helped her to regain her confidence and return to work. She then asked what I did in regard to CISM. I told her that I had trained her airline's team several years ago. Her response to this information was "Well, then, sir, you have touched my heart and I thank you for it."

It is important for us to listen carefully to those we serve with our CIRP or CISM teams. They know what helps them and they know when we have made a difference in their lives. In effect, each person represents a single case study that can help us to improve our crisis support service if we care enough to pay attention to their feedback and their advice.

References

American Psychiatric Association (1964), *First Aid for Psychological Reactions in Disasters.* (Washington, DC: American Psychiatric Association).
Antonellis, Paul, J., Jr. and Mitchell, S.G. (2005), *Posttraumatic stress Disorder in Firefighters: The calls that stick with you.* (Ellicott City, MD: Chevron Publishing).
Artiss, K. (1963), Human behaviour under stress: From combat to social psychiatry. *Military Medicine* 128, 1011-1015.
Bohl, N. (1991), The effectiveness of brief psychological interventions in police officers after critical incidents. In J.T. Reese and J. Horn, and C. Dunning (Eds.) *Critical Incidents in Policing, Revised* (pp.31-38). (Washington, DC: Department of Justice).
Bohl, N. (1995), Measuring the effectiveness of CISD. *Fire Engineering* 125-126.
Boscarino, J.A., Adams, R.E. and Figley, C.R. (2005), A Prospective Cohort Study of the Effectiveness of Employer-Sponsored Crisis Interventions after a Major Disaster. *International Journal of Emergency Mental Health* 7:1, 31-44.
Campfield, K. and Hills, A. (2001), Effect of timing of critical Incident Stress Debriefing (CISD) on posttraumatic symptoms. *Journal of Traumatic Stress* 14, 327-340.
Caplan, G. (1961), *An Approach to Community Mental Health.* (New York: Grune and Stratton).
Caplan, G. (1964), *Principles of Preventive Psychiatry.* (New York: Basic Books).
Caplan, G. (1969), Opportunities for school psychologists in the primary prevention of mental health disorders in children, In A. Bindman and A. Spiegel (Eds.) *Perspectives in Community Mental Health* (pp.420-436). (Chicago: Aldine).
Chemtob, C., Tomas, S., Law, W. and Cremniter, D. (1997), Post disaster psychosocial intervention. *American Journal of Psychiatry* 134, 415-417.
Deahl, M., Srinivasan, M., Jones, N., Thomas, J., Neblett, C. and Jolly, A. (2000), Preventing psychological trauma in soldiers. The role of operational stress training

and psychological debriefing. *British Journal of Medical Psychology* 73, 77-85.

Duffy, J. (1979), The role of CMHCs in airport disasters. *Technical Assistance Center Report* 2:1, 1; 7-9.

Everly, G.S., Jr. (1999), Emergency Mental Health: An Overview. *International Journal of Emergency Mental Health* 1, 3-7.

Everly, G.S., Jr. and Lating, J.M. (2004), *Personality-guided Therapy for Posttraumatic Stress Disorder.* (Washington, DC: American Psychological Association).

Everly, G.S., Jr. and Mitchell, J.T. (1997), *Critical Incident Stress Management: A new era and standard of care in crisis intervention.* (Ellicott City, MD: Chevron Publishing Corp).

Everly G.S., Jr. and Mitchell, J.T. (1998), *Assisting Individuals in Crisis: A Workbook.* (Ellicott City, MD: International Critical Incident Stress Foundation).

Everly, G.S., Jr. and Mitchell, J.T. (1999), *Critical Incident Stress Management: A new era and standard of care in crisis intervention, second edition.* (Ellicott City, MD: Chevron Publishing Corp).

Flannery, R.B. (2004), Posttraumatic Stress Disorder: the victims guide to healing and recovery (2nd ed.). (Ellicott City, MD: Chevron Publishing).

Forstenzer, A. (1980). Stress, the psychological scarring of air crash rescue personnel. *Firehouse* July, 50-52; 62.

Frederick, C.J. (Ed.) (1981), *Aircraft Accidents: Emergency Mental Health Problems.* (Washington, DC: National Institute of Mental Health, U.S. Department of Health and Human Services).

Freeman, K. (1979), CMHC responses to the Chicago and San Diego airplane disasters. *Technical Assistance Center Report* 2:1, 10-12.

Jenkins, S.R. (1996), Social support and debriefing efficacy among emergency medical workers after a mass shooting incident. *Journal of Social Behavior and Personality* 11, 447-492.

Kardiner A. and Spiegel, H. (1947), *War, Stress, and Neurotic Illness.* (New York: Hoeber).

Lindemann, E. (1944), Symptomatology and management of acute grief. *American Journal of Psychiatry* 101, 141-148.

Lindy, J. (1985), The trauma membrane and other clinical concepts derived from psychotherapeutic work with survivors of natural disasters. *Psychiatric Annals* 15, 153-160.

Mitchell, J.T. (1982), The psychological impact of the Air Florida 90 Disaster on fire-rescue, paramedic and police officer personnel. In R. Adams Cowley, S. Edelstein and M. Silverstein (Eds.), *Mass Casualties: a lessons learned approach, accidents, civil disorders natural disaster, terrorism.* (Washington, DC: Department of Transportation DOT HS 806302).

Mitchell, J.T. (2003), Major Misconceptions in Crisis Intervention. *International Journal of Emergency Mental Health* 5:4, 185-197.

Mitchell, J.T. (2004), Characteristics of Successful Early Intervention Programs. *International Journal of Emergency Mental Health* 6:4, 175-184.

Mitchell, J.T. (2005), *The Quick Series Guide to Stress Management for Emergency Personnel.* (Fort. Lauderdale, FL: Luxart Communications).

Mitchell, J.T. and Everly, G.S., Jr. (2001), *Critical Incident Stress Management: Basic Group Crisis Interventions.* (Ellicott City, MD: International Critical Incident Stress Foundation).

Mitchell, J.T. and Resnik, H.L.P. (1981), *Emergency Response to Crisis.* (Englewood Cliffs, NJ: Robert J. Brady Company, Subsidiary of Prentice Hall).

Mitchell, J.T. and Resnik, H.L.P. (1986), *Emergency Response to Crisis.* (Ellicott City, MD: Chevron Publishing (reprinted from original).

Neil, T., Oney, J., DiFonso, L., Thacker, B. and Reichart, W. (1974), *Emotional First Aid.* (Louisville: Kemper-Behavioral Science Associates).

Parad, H. J. (1971), Crisis Intervention. In R. Morris (Ed.) *Encyclopedia of Social Work*, vol. 1, 196-202.

Parad, L. and Parad, H. (1968), A study of crisis oriented planned short –term treatment: Part II. *Social Casework* 49, 418-426.

Richards, D. (2001), A field study of critical incident stress debriefing versus critical incident stress management. *Journal of Mental Health* 10, 351-362.

Roberts, A.R. (2000), An Overview of Crisis Theory and Crisis Intervention. In A. Roberts (Ed.) *Crisis Intervention Handbook: Assessment, Treatment, and Research.* (New York: Oxford University Press).

Roberts, A.R. (2005), Bridging the Past and Present to the Future of Crisis Intervention and Crisis Management. In A.R. Roberts (Ed.) *Crisis Intervention Handbook: Assessment, Treatment, Research.* (New York: Oxford University Press).

Salmon, T.S. (1919), War neuroses and their lesson. *New York Medical Journal* 108, 993-994.

Salby, A.E., Lieb, J. and Tancredi, L.R. (1975), *Handbook of Psychiatric Emergencies.* (New York: Medical Examination Publishing Company, Inc.).

Slaikeu, K.A. (1984), *Crisis intervention: A handbook for practice and research.* (Boston, MA: Allyn and Bacon, Inc.).

Solomon, Z. and Benbenishty, R. (1986), The role of proximity, immediacy, and expectancy in frontline treatment of combat stress reaction among Israelis in the Lebanon War. *American Journal of Psychiatry* 143, 613-617.

Stierlin, E. (1909), *Psycho-neuropathology as a result of a Mining Disaster March 10, 1906.* (Zurich: University of Zurich).

Vogt, J., Leonhardt, J., Köper, B. and Pennig, S. (2004), Economic Evaluation of CISM – A Pilot Study. *International Journal of Emergency Mental Health* 6:4, 185-196.

Wee, D.F., Mills, D.M. and Koelher, G. (1999), The effects of Critical Incident Stress Debriefing on emergency medical services personnel following the Los Angeles civil disturbance. *International Journal of Emergency Mental Health* 1, 33-38.

Chapter 4

Critical Incident, Critical Incident Stress, Post Traumatic Stress Disorder – Definitions and Underlying Neurobiological Processes

Jörg Leonhardt and Joachim Vogt

The aim of this chapter is to give definitions and metaphorical illustrations of the terms Critical Incident (CI) and Critical Incident Stress Reactions (CIS). Moreover, the neurobiological processes involved in stress reactions are briefly described. Finally, the Post Traumatic Stress Disorder (PTSD) is introduced as the consequence of CIS reactions that were not sufficiently coped with.

Critical Incident (CI)

Crucial aspects in dealing with Critical Incident Stress Management (CISM) are the description and the definition of the Critical Incident (CI) itself. The definition of the CI provides the basis for understanding the crisis, the Critical Incident and CISM. The distinct and clear definition is a premise of defining and constructing all further CISM modules. A distinct and clear definition enables participants to offer, apply and accept CISM.

The situation of a CI goes beyond normal, common experiences and, therefore, the people affected find themselves in an extremely stressful condition. Usually, those affected experience unknown or exceptional reactions called Critical Incident Stress Reactions (CIS).

Therefore, the definition of a CI is as follows:

Every situation experienced by a person that causes exceptional or unusual reactions.

The key issues of this definition are: *every situation* and *exceptional or unusual reactions*.

Initially, it might be tempting to define the incident itself when giving a definition of the CI relating to situations such as incident of major loss, catastrophe, threat of life, injury, death and so on.

This approach makes sense as the above mentioned situations will most probably cause critical incident stress reactions and will be judged as catastrophes by most people. However, a definition based on the incident neglects the individual: the people affected with their individual reactions. Moreover, a definition understood in this way would exclude situations without the above characteristics, but which can have the potential to cause critical incident stress reactions for some people.

In aviation, especially in air traffic control (ATC), critical situations like, for example, a separation infringement between two aircraft does not usually result in an accident. Air traffic controllers have to maintain safe separation between aircraft. An infringement of these separation minima does not usually lead to an accident and is normally not noticed by the public. However, it can cause Critical Incident Stress Reactions in the air traffic controller involved.

In this case, a peer model of crisis intervention, in which other air traffic controllers conduct the CISM interventions, is especially important, because they as a member of this professional group can realize and comprehend the impact of the separation loss (see Chapter 7). The professional peer understands the critical incident stress reactions of the colleague. This cooperative understanding is necessary to calm the reactions down and support the individual coping process.

In any kind of crisis intervention, it should always be the people and their individual reactions that are the focus and that define the treatment: "Crisis is in the eye of the beholder" (Jeffrey Mitchell, personal communication).

From the experience gathered in CISM interventions, it is known that different people experience critical incidents in different ways. The different experiences lead to different reactions. Thus, a group of people experiencing the same incident may reveal patterns of critical incident stress reactions of different qualities and intensities. It is essential for the definition of a critical incident to start from the viewpoint of the people affected and secondly to consider the incident itself.

People usually can be characterized by different psychophysical *states of charge* (see Figure 4.1) throughout life, bearing in mind that this status is not constant – it changes from time to time. Using the metaphor of a battery, we can imagine that the capacity decreases with major demands and little time to recharge. The battery still provides energy; however, the flashlight does not shine as brightly, the walkman plays music slowly and the battery power is fading.

In a figurative sense, a person who uses most of their capacities to deal with everyday hassles and who has little time to recharge the *batteries* for a healthy stress management is likely to act differently in a critical situation than someone with a *fully charged battery* – with a viable psychophysical status.

Factors like trouble in the job, separation or divorce, illness, depression or other impairments use up mental energy, especially if the problems are not coped with sufficiently. Additionally, in a stressful situation with high mental demands there is usually little time for recreational activities to recharge.

Conversely, the *batteries* are in top condition in situations with low stress, good relations with family and friends and an intact and healthy life. In such a psychophysical status, the people's own resources suffice to handle the events,

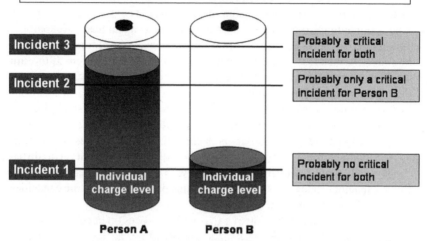

Figure 4.1 Different critical incidents (CIs) meet person A in a good and person B in a limited psychophysical condition

even more so if the people affected find support within themselves and the social environment. This social support further facilitates one's own personal coping strategies (see Chapter 12).

Figure 4.1 exemplifies different states of charge meeting different CIs. Figure 4.1 shows that person A can be characterized as being in a good physical and mental condition, the social situation is optimal; life is like it should be. In our metaphor, a fully charged battery (90 per cent) can be relied on.

Person B is in an extremely stressful situation, facing illness, separation, social and/or work-related problems. These factors are demanding on the mental coping capacities and leave only a little capacity for additional events (30 per cent). Person A and B have different coping capabilities in the face of a CI and thus show different reactions.

Both, person A and person B will probably cope with minor incidents demanding less than 20 per cent of their respective coping capacity (incident 1 in Figure 4.1). Both will probably experience an incident demanding 100 per cent coping capacity as critical (incident 3 in Figure 4.1). However, person A is likely to cope with a CI exposing to a stress level of 70 per cent without major impairments, whereas person B will probably experience severe CIS reactions (incident 2 in Figure 4.1). Facing different incidents, person A is likely to handle a CI with a demand on his/her coping capacities of 70 per cent, whereas person B is likely to experience CIS already facing a CI demanding 30 per cent of his/her coping capacities.

Applying the event-related definition of a CI would result in further complications, as an extremely stressful CI leads to minor impairments of person A, while a less stressful situation provokes an *overreaction* of person B. This supposed discrepancy might cause a lack of understanding. An *objective* assessment from an external perspective without considering the current individual *state of charge* of the person affected is prone to misjudgments. These misjudgments again lead to inappropriate support: more than necessary for person A, less than necessary for person B.

Moreover, person B is probably not open for a CISM intervention, as the strong personal reactions are discrepant to the social environment's assessment of a minor incident, which can lead to self-doubts.

Important note: In this metaphor we have to consider that person A may cope alone with a critical incident; however, a lot of psychophysical energy is used by this coping process and the *battery status* is decreased. This underlines the frequently changing psychophysical status. Person A is still person A, but in a different psychophysical balance, which means coping with a critical incident using merely their own capacities today does not mean that this person has the same capacities to cope with a less severe incident next week.

The above mentioned definition of a CI – with the focus on uncommon reactions of the people affected – considers a person's *state of charge* and the subsequent coping capacity. The focus lies upon the person affected and the respective reactions.

By focusing on the reactions of the affected person, peers as well as people outside ATC can regard the reactions as normal in the face of an abnormal situation. This facilitates the offering and acceptance of support. The uncommon reactions serve as parameters for the assessment of appropriate support. Uncommon in this sense means: different from otherwise, longer and more intensive. Even though the definition focuses on the person affected, under these particular circumstances this individuality contributes to objectifying the individual stress reaction. Moreover, the feeling of uniqueness is eliminated. Thoughts or comments like "why me" or "I am the only one, others are not affected" give way to a more realistic and objective assessment of the subjective reactions like "others would feel the same" or "in my current situation it is too much for me, at other times I could deal with it better".

Objectifying is supported by integrating the current life situation of the affected person, thus making clear that the CIS reactions are not considered a static state. Considering the circumstances – in our metaphor the poor state of charge – coping capabilities are lower than they used to be. The affected person realizes that different situations can be coped with differently and can tie in with coping strategies and resources of past stress situations.

Due to the reasons mentioned above, the definition of a CI should always refer to the individual effects and reactions to a CI. Nevertheless, the severity and the extent of the incident are not to be neglected. Especially when dealing with major incidents (like natural disasters, incident of major loss, terrorist attacks), appropriate measures should always be taken as secondary prevention. The five issues of the strategic formula, namely target, type, timing, theme and team have to be considered for the crisis intervention plan (see Chapter 3). These measures can comprise all interventions of the CISM program; however, one should not be tempted to try to objectively assess the individual reactions.

Critical Incident Stress Reactions (CIS)

CIS emerge from a Critical Incident and their intensity depends on the psychophysical status of the responding person. They can occur on four levels (see Chapter 3):

- Cognitive
- Emotional
- Physical
- Behavioral.

The definition of CIS is: *The psychological and physical reactions and behavioral changes which a person experiences after a CI. These reactions are normal reactions to an abnormal event.*
Reactions on the different levels are for example:

Cognitive:
- General confusion
- Difficulty in decision making
- Difficulty in identifying people known to the individual
- Disorientation in terms of time and place
- Change in readiness to react to situations
- Changed perception of surroundings
- Distrust
- Nightmares
- Deterioration in ability to concentrate and being alert
- Memory lapses, blanks.

Emotional:
- Fear and insecurity
- Feelings of guilt
- Feeling of being overwhelmed/helplessness
- Anxiety
- Irritability/aggression
- Fits of anger
- Increased excitability
- Panic attacks
- Over exaggerated expressions of grief
- Suppression of feelings/elusive behavior
- Lack of emotion or outbreaks of emotions
- Depression.

Physical:
- Sudden dizziness/feeling of faintness
- Dizziness/Numbness
- Sleeping disorder
- Faster pulse or higher blood pressure
- Breathing difficulties
- Dimness of vision
- Chills and fever
- Teeth grinding
- Increased fluid intake
- Drowsiness
- Nausea and vomiting
- Muscle twitching/nervous twitching/paralysis
- Headaches and chest pains
- Shock.

Behavioral:
- Disturbed appetite
- Speech changes
- Changes in social behavior
- Isolation
- Over sensibility
- Hurry, restlessness
- Uncontrollable movements (for example ticks)
- Increased substance use
- Anger expression
- Increased excitability
- Panic attacks
- Over exaggerated expressions of grief
- Suppression of feelings/elusive behavior
- Lack of emotion or outbreaks of emotions
- Retreat, immobility, hyper mobility.

CIS can occur with different combinations, over a longer period of time, in turn or delayed.

In order to assess the individual and work-related impacts of CIS, it is important to consider that physical reactions are more easily detectable than cognitive reactions. On the one hand, this can lead to the mere medical treatment of bodily symptoms ignoring that they are CI-related and getting in the way of psychological coping. On the other hand, losses in cognitive capacity – in our metaphor a poor state of charge – are usually first recognized when needed the next time. For an air traffic controller or a pilot it is essential to make quick and clear decisions; precise information intake, processing and action are the backbones of this work. CIS can impair these vital abilities – in the worst case first noticed when cognitive capacity is next needed.

Therefore, CISM interventions are also preventive in the sense that the reflection of the CI and CIS may help the affected person to detect latent cognitive impairments.

Note: As will be reported in Chapter 12, it is of utmost importance that supervisors recognize that after a critical incident cognitive ability may be latently limited. It is the responsibility of the supervisor to act accordingly by, for example, taking the controller away from their position and asking for a peer.

Neurobiological Aspects of Critical Incident Stress Reactions

In order to understand why Critical Incident Stress Reactions are considered as normal reactions to an abnormal event, it might be helpful to give an insight into the underlying neurobiological mechanisms. Hans Selye (1936, 1950) described very early certain stress reactions as a general adaptation syndrome to all kinds of stressful situations.

In a situation perceived as threatening, uncommon, extreme or dangerous, the oldest part of our brain facilitates an archaic reaction. This reaction sets free an enormous amount of neural and chemical messages, which alert the body in a split second in order to be able to react at once.

The biological reactions are usually unconscious and automatic; the biological system reacts to a great extent automatically, thorough cognitive assessment or planning is usually unfeasible.

The body is able to act immediately without even thinking or deciding what to do. The reactions are similar to a reflex, like ducking down, raising the arm to protect the body or turning the head for orientation without thinking.

As an illustration, we imagine someone is driving and suddenly a critical situation occurs requiring, for example, braking sharply on the highway. In a split of a second, we react and try to handle the situation, to avoid the crash. That means we steer, break, move the arms and feet in a reflex action without being conscious about this. This reaction has to occur immediately as all cognitive processes would take too long and we would not be able to handle the situation anymore. Afterwards we are aware of our stress responses, like a quick pulse, being weak in the knees, having a nervous tremor, we are excited and start to comprehend what has just happened.

The biological processes are normal in abnormal situations. The limbic system, a part of the brain emerging very early in phylogeny, and in humans located between the brain stem and the neocortex, has taken over control and aroused the body to full alert. The mentioned after effects can be traced back to the neurobiochemical messages that have been sent. Trauma and brain research have paid special attention to the limbic system. The limbic system is also involved in the control of our survival behavior like eating, drinking, sexuality and the so called fight or flight reactions (Cannon 1914, see also Chapter 7). Moreover, it is directly connected to the autonomous nervous system and therefore plays its part in the regulation of muscles, the heart and breathing rate, functioning of the lungs, blood pressure and so on.

Two nervous systems control these functions: the sympathetic and the parasympathetic nervous systems. The sympathetic nervous system is activated in stressful situations irrespective of the connotation: The individual may experience the situation as positive in the sense of, for example, a challenging work task or sexual stimulation; for these cases the term eu-stress was coined (eu being Greek for good). The term dis-stress (dys being Greek for bad) describes negatively experienced stress, for example fear-inducing, situations.

A parasympathetic tone facilitates, for example, digestion and relaxation. The limbic system reacts to threats by releasing hormones, which again cause reactions of the sympathetic nervous system, and set free the adrenocorticotropic-releasing hormone.

This reaction is introduced via two pathways of the limbic system: the amygdala and the hypothalamus. The body is prepared to fight or flight immediately. Diverse bodily reactions occur:

- Elevated blood pressure for a better blood circulation through the muscles
- Increased heart and breathing rate to process more oxygen
- Constriction of the blood vessels
- Larger pupils for a larger visual field.

At the same time, a slower reaction chain, the hypothalamic pituitary adrenal axis (HPA) releases the anti-hormone cortisol which stops the alarm reaction (that is the reaction of the sympathetic nervous system). The body is put into a homeostatic status again. This is a parasympathetic reaction balancing the sympathetic reactions described above.

Table 4.1 gives an overview of how situations are perceived, which brain parts are involved, which responses are most frequently shown, which hormones are released and which diseases are connected to overdriving one of the systems.

The fact that a single region of the brain automatically takes control when threatened in order to prepare the body for fight or flight, in combination with the automatic initialization via neurobiochemical processes, makes it clear why CIS have to be considered as a normal response to an abnormal event. We are simply not able to react any differently.

Post Traumatic Stress Disorder (PTSD)

While critical incident stress reactions are a normal survival mechanism, Post Traumatic Stress Disorder (PTSD, see also Chapter 3) is the pathogen manifestation of CIS which were not sufficiently coped with. Chronic CIS become apparent in lasting behavioral and psychophysiological changes. High blood pressure, sleep disorders, nervousness, and déjà-vu sensations might occur. The latter can impair performance and well-being as intrusive memories. Behavioral changes manifest themselves in, for example, unsociability, social isolation and depression.

Table 4.1 The Selye (1936, 1950) distress and the Cannon (1914) eu-stress axis in an integrative model (modified according to Henry and Stephens 1977, p. 119)

Perception	Threat to control (Eu-Stress)	Loss of control (Dis-Stress)
Brain area	Amygdala Behavioral arousal with challenge to status and desiderata	Hippocampus-Septum Inhibition of spatially organized behaviors and status
Response	Defense/attack Territorial control Mobility Aggression	Conservation/withdrawal Restricted mobility Subordination Low drives
Response axis	Cannon Sympathetic adrenal-medullary system	Selye Pituitary adrenal-cortical system
Hormones	Noradrenaline ↑ Adrenaline ↑ Cortisol ↔ Testosterone ↑	Adrenocorticotropic hormone ↑ Adrenaline ↔ Cortisol ↑ Testosterone ↓
Body responses	Heart rate ↑ Blood pressure ↑ Respiration ↑ Glucose ↑ Blood coagulation ↑ Muscle tone ↑ Gastrointestines ↓ Kidneys ↓	Protein decomposition ↑ Liver glycogen storages ↑ Immune system ↓ Gonads and fertility ↓
Long-term diseases	Cardiovascular Gastrointestinal	Infectious Depressive

Key: ↔: no change, ↑: increase, ↓: decrease

The current issue of the International Classification of Diseases (ICD10) of the World Health Organization (WHO) defines PTSD as a delayed response to critical, stressful incidents. These usually involve extraordinary threat or catastrophic dimensions which would cause deep despair in any person. Examples are situations where someone's life has been threatened by violence or being diagnosed with a life-threatening illness or a severe injury has occurred (for example, to the victim or a witness of physical abuse, violence in the home or in the community, automobile accidents, disasters such as flood, fire, earthquakes). In contrast to the immediate biological stress responses described above, PTSD can occur with a latency of a few weeks up to several months after the CI. The symptoms of PTSD may last from several months to many years. Certain predispositions like, for example, obsessive-compulsive personality can promote the development of PTSD or intensify the symptoms. Typical symptoms are intrusive memories (déjà-vu, flashbacks) and nightmares that occur on the basis of emotional numbness and experienced helplessness. PTSD patients often show apathy towards other people and the environment, they are usually cheerless and inactive. They especially avoid situations that could remind them of the CI. Anxiety and depression are the emotional concomitants, suicidal thoughts can also occur.

In contrast to the behavioral passivity, the physiological systems are usually over aroused in connection with undue vigilance, startled responses, and sleep disturbance.

PTSD progresses in waves but, in the majority of cases, a sustainable recovery can be achieved given the appropriate treatment. Once the trauma has occurred, early intervention is essential. The best approach, however, is the prevention of trauma. In order to prevent PTSD, CISM has proved to be the method of choice. As outlined above and further to be elaborated in the following chapter, support from CISM-trained peers is the most effective, instantaneous, and natural application in aviation.

Conclusion/Summary

A Critical Incident has been defined as every situation experienced by someone that causes exceptional or unusual reactions. These psychological and physical reactions are initiated by neurobiological pathways and can be considered normal reactions to an abnormal event. In aviation, Critical Incidents can occur that do not lead to accidents and which, thus, are not even perceived by the general public. The professionals involved, however, can show Critical Incident Stress Reactions. Qualified colleagues (peers) can comprehend the impacts of such Critical Incidents and offer support to their affected colleagues. A Critical Incident Stress Management Program based on this peer support can successfully prevent work-related as well as individual short- and long-term consequences, for example losses of cognitive capacity or Post Traumatic Stress Disorder (see Chapter 12).

References

Cannon, W.B. (1914), The emergency function of the adrenal medulla in pain and major emotions. *American Journal of Physiology* 33, 356-372.

Henry, J.P. and Stephens, P.M. (1977), *Stress, health and the social environment*. (Berlin: Springer).

Selye, H. (1936), A syndrome produced by diverse nocuous agents. *Nature* 138, 423-7.

Selye, H. (1950), *The physiology and pathology of exposure to stress*. (Montreal: Acta Inc.).

Chapter 5

Critical Incident Stress Management Intervention Methods

Jörg Leonhardt

This chapter describes the different methods and parts of the CISM model, the different types of intervention, their time structure and function within the whole system.

Crisis intervention is a necessary and supporting measure, which has been tailor-made for people in crises. Crises or critical situations in aviation have very different characteristics and a wide range of consequences (see Chapter 3). However, a common feature of all critical situations is that the people concerned (victims, survivors, professionals involved) need support to better cope with their stress reactions connected to the crisis. "Crisis Intervention was developed in response to the acute mental health needs of those in crisis" (Mitchell and Everly 2001).

The goal of crisis intervention is the reduction of stress reactions and the avoidance of a Post Traumatic Stress Disorder (PTSD). Moreover, the individual ability to act is restored and the coping strategies are reactivated. If necessary, the person affected is referred to further treatment after the crisis intervention.

The crisis intervention does not signify psychotherapy or a supplement of such. Crisis intervention is direct support immediately after the Critical Incident. The crisis intervention focuses on the incident itself and the connected stress reactions – of the acute and prevailing crisis – and does not focus on the treatment and discussion of past incidents or impairments.

Crisis intervention is characterized by (see Chapter 3):

- Immediacy – immediate support
- Proximity – proximity to the incident regarding time and place
- Expectancy – association between intervention and expectancies (for example reduction of stress reactions).

The crisis intervention should be applied directly and without delay at the work place or the scene and conclude with clear and feasible next steps. The crisis intervention can be considered an antidote for the acute psychological disturbances caused by the crisis. The antidotes are:

- Structure as an antidote to chaos
- Cognition as an antidote to excessive emotions

- Catharsis and disclosure to release psychological tension
- Understanding as an antidote to loss of control
- Action as an antidote to helplessness.

Research has shown that human-made disasters are more psychologically pathogenic than are natural disasters. Terrorism may be the most pathogenic of all due to its unpredictable and unrestrained nature (Mitchell and Everly 2001).

According to this quotation and considering the disasters that have happened in aviation or could happen, a need for CISM is clearly given. Aviation employees like airline, airport or ATC staff have a professional role in the process that may result in an accident. They are in danger themselves and/or are responsible for the safe conduct of flight. This personal involvement and incident-related feelings of guilt are key risks for potentially traumatizing events.

Crisis intervention – that is Critical Incident Stress Management – should always be conducted immediately or as early as possible after the incident.

There is a strong argument for providing acute psychological first-aid as early as practical following a traumatic event (Bisson, McFarlane and Rose 2000, pp.39-59).

CISM is one component of a spectrum of care. The whole spectrum comprises social counseling, psychotherapy, Employee Assistance Program (EAP), rehabilitation, counseling and treatment by physicians or psychologists as well as other measures. The crisis intervention is always the first step of this whole spectrum,

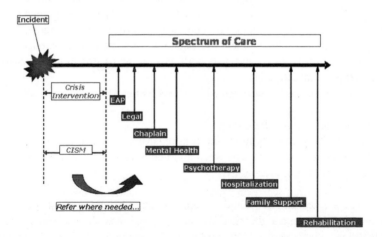

Figure 5.1 Spectrum of care: CISM as the first element in psychosocial help (modified according to Mitchell and Everly 2002); EAP: Employee Assistance Program

as the intervention restores the stability of a person and therefore the structure needed for further treatment (Figure 5.1).

CISM is a comprehensive, systematic and multi-component approach for the management leader of critical incident stress. Such an approach should be based on intertwined measures that can be adapted to the special situation and to the needs of the people. CISM requires substantiated and evaluated training. In the following, the basic elements of a CISM program will be explored.

The basic elements of a CISM program are:

- Information and Training
- Pre-incident Preparation and Education
- On-scene Support Services
- One-on-one Crisis Intervention/SAFER-R
- Demobilization
- Crisis Management Briefing/CMB
- Defusing
- Critical Incident Stress Debriefing/CISD
- Family Support
- Organizational/Community Support
- Referrals.

Information and Training

Informing all employees and managers and training the managers and peers are the basic elements of a comprehensive Critical Incident Stress Management Program. Information must be provided about the program, its objectives and type of effects, organizational rules and procedures. Moreover, management and peers have to be informed about critical incidents and Critical Incident Stress Reactions, how they can be identified and assessed. The training comprises the instruction of the peers and is based on three interacting courses:

- Individual Crisis Intervention and Peer support
- Group Crisis Intervention
- Advanced Group Crisis Intervention.

The training and the trainers should be certified and accredited by an internationally recognized organization like the International Critical Incident Stress Foundation ICISF (see Chapter 11). The CISM program manager and the crisis intervention team (CIT) leader should also be trained in crisis intervention and strategic planning. Especially the first management level, such as operational managers or supervisors (see Chapter 12), needs CISM-related training facilitating the first steps after a critical incident, that is assessing the ability of the employee concerned to work and, if need be, replacing the employee and offering a CISM intervention (see Chapter

4). Moreover, the management should have a comprehensive and well-known policy regarding the company's CISM program. Upper management support is crucial to the functioning and acceptance of the CISM program (see Chapter 2).

Pre-incident Preparation and Education

Pre-incident preparation and education means preparing for the CISM measures. If peers are dispatched to a crisis area, for example, it has proven helpful to mentally prepare the staff for this special situation. General information on stress and trauma are refreshed. Additionally, the peers are mentally and physically trained for the expected particularities of the situation. Moreover, acute as well as long-term coping strategies are introduced. The aim is to minimize the *horror of the unexpected*, to be prepared for possible mental and physical reactions and to keep one's own resources available in the crisis.

One-on-one Individual Crisis Intervention

Most of the crisis interventions happen individually with face-to-face communication, in exceptional cases via the telephone. The most common form of the individual crisis intervention is the SAFER-R model.

*S*tabilize	Stabilizing the situation and stopping the influence of the immediate sensations (smell, sight, taste, hearing).
*A*cknowledge	Acknowledging the crisis by inquiring about the facts, the reactions to the event and the possibility to talk about the situation and the reactions.
*F*acilitate	Central in this step is the normalization – the responses which are experienced as unusual are normal in the unusual situation. To facilitate the understanding that these reactions are normal in the abnormal situation is one mean of normalization.
*E*ncourage	Encouraging the adaptive coping functions.
*R*ecovery	It is evident that the person in crisis is on his/her way to recover and restore her/his normal functions and abilities.
*R*eferral	If needed, referral to other professional help (Spectrum of Care, Figure 5.1).

The SAFER-R model is applied by peers and carried out immediately after the Critical Incident.

Demobilization

Demobilization is to be understood as the opposite of mobilization. Demobilization is a measure for large groups at the end of an operation. The operation is over, the equipment has been laid down and the people have come together. At first, basic needs are cared for (food, drinks, warming up and so on). Afterwards, the participants are informed about stress and stress management. If required, one-on-one interventions are initiated and phone numbers of the peers are distributed.

The demobilization is a passive event with an information phase (approximately ten minutes) and with time for rest, food and drink (approximately 20 minutes). The demobilization does not substitute necessary CISM measures like defusing or debriefing. However, the demobilization can be used to identify the necessity of further CISM measures.

Crisis Management Briefing (CMB)

CMB is one of the more recently developed CISM intervention measures. CMB is also a measure for larger groups. However, it has been conceived for incidents affecting a larger group of people.

> Structured large group community/organizational town meetings designed to provide information about the incident, control rumors, educate about symptoms of distress, inform about basic stress management and identify resources available for continued support, if desired. Especially useful in response to violence/terrorism. (Mitchell and Everly 2002).

CMB should be operated in cooperation with a representative of the organization or community and a CISM team member. The group should be homogenous and the time for discussion and questions should be restricted. Information on the incident and the status quo are passed on as well as information on stress, stress management and the respective support. A list of the phone numbers of CISM team members is handed out. CMB is to be repeated regularly for a fixed period of time. By doing so, up-to-date information is ensured. Even the information *no news* is important to people in a crisis.

CMB has proven very effective for dealing with the disaster in Überlingen within the German air navigation service provider DFS Deutsche Flugsicherung GmbH (see Chapter 10). The employees were informed regularly about the latest news and about CISM over the period of one week. The information was passed on by one manager and one CISM peer.

Defusing

Defusing is one CISM measure for smaller groups (less than 20 people). Defusing takes place on the day of the incident or the day after. Usually, it takes less than one hour.

The aims of defusing – as with all CISM measures – are the reduction of stress reactions and the re-storing of adaptive functions. Defusing needs a CISM team of at least two members. A mental health professional (MHP) is not required. However, a qualified peer has to run the defusing. One team member is in charge and leads the conversation.

Defusing consists of three phases:

1. Introduction
2. Exploration
3. Information.

Defusing starts on the cognitive level, continues on an affective level, returns to and ends with cognition (see Figure 5.2). This structure is important for the effectiveness of defusing because people in crisis need, firstly, easy access to the facts (cognitive level) and then, secondly, a guided emotional process (affective level), before, thirdly, they are able to return and further pursue their own coping strategies (cognitive level).

For these reasons, the intervention process has to follow the structure. Ending on the cognitive level, in particular, facilitates further planning and coping.

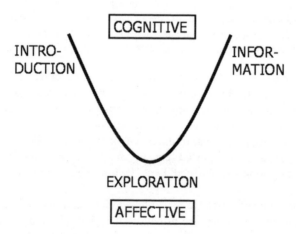

Figure 5.2 Three phases of defusing (modified according to Mitchell and Everly 2002)

In most of the cases, defusing is followed by a debriefing.

If a defusing is missed, or if it is impossible to provide due to circumstances beyond the control of the CISM team, then the efforts to provide the defusing should be abandoned and the team should prepare to provide a CISD when appropriate (Mitchell and Everly 2002).

Critical Incident Stress Debriefing (CISD)

The CISD – Critical Incident Stress Debriefing – referred to in this book is a structured model for crisis intervention in groups consisting of seven phases. The CISD has been developed and validated by Jeffrey T. Mitchell and George S. Everly. Also, it has proven very helpful and efficient in many different field operations (for example fire fighters, rescue services, airlines, air traffic control, airports and so on).

Debriefing is a CISM measure for groups of up to 20 people. A mental health professional is in charge and supported by four to five peers (see Chapter 3). Ideally, debriefing is run within two to five days after the incident. The debriefing takes two to three hours.

The peers should have the same profession as the people affected in order to be more able to assess the reactions and symptoms. In this way, the aspect of normalization, as a central issue of crisis intervention, can be more authentically translated into action. Moreover, the group should be as homogeneous as possible and should have been equally exposed to the traumatic event.

The seven phases of the debriefing are as follows (Figure 5.3):

1. Introduction
2. Fact
3. Thought
4. Reaction
5. Symptom
6. Teaching
7. Re-entry.

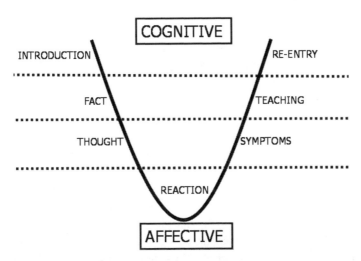

Figure 5.3 Seven phases of debriefing (modified according to Mitchell and Everly 2002)

As with the defusing process, the intervention starts on a cognitive level, gradually turns to the emotional level and later enters the cognitive level again. At the end of a debriefing, individual coping strategies should have been defined and the participants should have developed concrete further steps to take.

The structure of the debriefing process follows the same pattern as described for the defusing process: from a cognitive level through an emotional catharsis back to a cognitive level. The structure is important for the same reasons, and for the CISM facilitator it is essential to know on which level (cognitive/affective) the discussion is taking place during the process (see Figure 5.4).

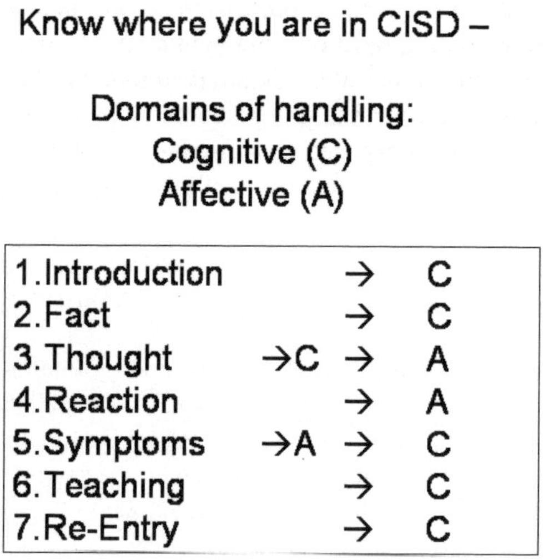

Know where you are in CISD –

Domains of handling:
Cognitive (C)
Affective (A)

1. Introduction	→	C
2. Fact	→	C
3. Thought	→C →	A
4. Reaction	→	A
5. Symptoms	→A →	C
6. Teaching	→	C
7. Re-Entry	→	C

Figure 5.4 Know where you are in Critical Incident Stress Debriefing (CISD) (modified according to Mitchell and Everly 2002)

Chapter 10 illustrates the debriefing process with the standard procedure as used after the Überlingen accident. It describes how the people involved, that is air traffic controllers (ATCOs), peers, and the mental health professionals (MHP), experienced the procedure and its effectiveness from their different perspectives.

Family Support

The person affected is likely to pass her/his intensive impressions and emotions on to family members. On the one hand, the person affected is likely to tell stories about the incident. On the other hand, changed reactions and changed behavior will be noticed by family members and thus the impressions and emotions transfer to these people as well.

Both aspects might make family members feel insecure and helpless, which often leads to further difficulties. That is why supporting the family is one major element of a comprehensive CISM program. The supportive measures might be information and counseling or advice. The measures have to be chosen and adapted to the special needs of the different family members considering, for example, their age and educational status. Children need a different kind of support than teenagers, and a partner needs a different kind of support than the parents. The aim of the measure is to always maintain the family's ability to act and to make the crisis-induced changes understandable. Only specially trained CISM staff should provide family support. The purpose is not to replace the support provided by other people close to the family; on the contrary, an integration of their support makes perfect sense.

Organizational/Community Support

In terms of prevention, this kind of support deals with the preparation and the establishment of a CISM program within organizations and the community and with the integration of such a program in the local emergency plan.

In an acute crisis, community support might be the direct support with CISM measures. There are many examples where organizational/community support is needed: violence in schools, natural disasters, local incidents or cases in which a community is particularly affected by the death of community members in an accident outside the community.

Referrals

During a CISM intervention, the need for further help may be identified. In this case, a referral to professional services is made. It is important that these services are known and referred to by the CISM facilitator. One aim of the peer training is to recognize the need for further assistance and initiate the standard procedure for a referral according to organizational practice.

So far this chapter has described the CISM methods. During all CISM applications, normalization and the role of the peers are central issues. Due attention shall be paid to these two issues in the following sections.

Normalization

As previously mentioned, normalization is the crucial aspect of a crisis intervention for groups or for a single person. Normalization means supporting the person affected, accepting the reactions perceived as unusual and learning that they are normal reactions to abnormal events. If someone has experienced a terrible situation, it is normal not to sleep well, to have an increased pulse, to experience fear and worries and other reactions. For the people affected, these reactions seem abnormal and unusual at first. Considering the context of the incident, however, they make sense and are comprehensible (Chapter 4). It is very helpful to realize that these reactions make sense in the context of the incident in order to be able to integrate them into one's self-image and finally to cope with the reactions.

To give an example: If a child wakes up in the morning and finds the pet, for example the hamster, dead in its cage, this would be a critical incident for the child. The child will be sad, feel insecure or even distorted. The center of the child's world will be the dead hamster.

At the same time, the child experiences uncommon reactions. It is important to communicate that it is normal and comprehensible. Sometimes giving an example of a similar situation one has experienced helps to normalize the reactions.

When children accept their own reactions as normal, they will be able to integrate the reactions and therefore regain the ability to act. The mourning will go on for a while; however, it will not lead to desperation. Moreover, children can actively deal with the situation, for example bury the hamster or perform other rituals. Pain and grief are not suppressed, but integrated and therefore a part of the incident, which leads to sustained coping. In the long term, coping capabilities after critical incidents will be higher.

The same happens when ATCOs or pilots as peers tell the affected colleague that they can understand the reaction, that they would have done the same or something similar in a situation like that. This is the outstanding advantage of the peer model in general and in ATC, particularly with respect to the special requirements (see Chapter 7).

Normalization mechanisms are similar in every context. The feeling of being understood and knowing that others would have similar feelings reduces the stress reactions and integrates them into the individual and professional self-image.

Normalization is a powerful linguistic tool; however the words have to be chosen carefully. It is important to normalize the reactions and not the situation. A peer can say for example "I can understand your reactions, other controllers would react similarly" thus normalizing the reaction. Trying to normalize the situation in saying "a separation infringement can occur in ATC, this could happen to other controllers as well" is not the meaning of normalization. The effects of such an approach can be harmful and can increase the feeling that something is wrong with the person. "If such a situation is normal in our profession, why am I reacting so intensely?"

Using examples of different life situations can be helpful to illustrate the underlying coping mechanism. In our example with the hamster, it would not at all be helpful saying that it is normal for a hamster to die after a couple of years. This would not reduce the reactions – grief – of the child; on the contrary it would keep the child from integrating the own reactions.

Normalization and crisis communication techniques are essential in peer training (see Chapter 11).

Role of the Peers

Peers, colleagues trained in CISM, play the most active and most effective role in the CISM program. Who else can understand how someone feels after a job-related critical incident if not a colleague? Who would you rather believe in a crisis that the reactions are understandable or even normal if not a colleague? And who could better assess the reactions and behavioral changes of a colleague than a trained peer? Peers are the backbone of any CISM program, especially in aviation (see Chapter 7).

Peers should be motivated and experienced, be trusted by their colleagues and be empathetic. Peers should be provided with adequate training (see Chapter 11), regular supervision and refresher courses.

Management must ensure this continuous qualification of the peers. Support and appreciation of the CISM program and the peers' commitment by management plays a vital role in the success of the program and the continuous motivation of the peers (see Chapter 2).

References

Bisson, J.I., McFarlane, A. and Rose, S. (2000), Psychological debriefing. In E. Foa, A. McFarlane and M. Friedman (Eds), *Effective Treatments for PTSD* (pp.39-59). (NY: Guilford).

Mitchell, J.T. and Everly, G.G. (2001), *Critical Incident Stress Debriefing. An Operations Manual for the Prevention of Traumatic Stress among Emergency Services and Disaster Workers*. (Ellicott City, MD, USA: Chevron Publishing).

Mitchell, J.T. and Everly G.S. (2002), *Training material for CISM Courses*.

Chapter 6

A Brief History of Crisis Intervention Support Services in Aviation

Jeffrey T. Mitchell

There is a pressing need to commit to writing a history of crisis intervention support services in the aviation industry. Future generations of aviation personnel may need to know the theoretical and historical roots of their services to their aviation colleagues. Others, who are just beginning to formulate crisis response teams, will need to know the paths that others have walked before them. History and a strong theoretical foundation provide a firm base from which current programs and services can be confidently provided. Good theory, in essence, leads to good practice.

Writing a history of crisis support services in the aviation industry is, at best, a challenging task. Absolute historical comprehensiveness is a virtual impossibility and will not be attempted here. Only the historical highlights that can be accurately supported by the author's memory and by available documents will be addressed in this chapter. It is likely that some situations and some individuals were inadvertently left out of this history. Those omissions are the result of too long a passage of time from the occurrence of the events until this opportunity to write a history.

There are simply too many airlines and too wide a range of aviation services including air traffic control systems in the world to cover completely their involvement in Critical Incident Stress Management in this chapter. Attempting to do so would make the chapter so large and so unwieldy that it would be unreadable. Many of the air carriers and aviation ancillary services that have organized crisis support teams started their programs at different times over the last 20 years. Each airline and air traffic control system initiated their programs under vastly different circumstances and each has its own history of their support services. Some crisis support teams developed quietly and did not announce the team's existence to anyone other than the employees within their companies. It is, therefore, impossible to provide an accurate record of each historical development which includes each individual's and each organization's contribution.

There is no doubt, however, that both the airlines and the ancillary aviation services have many dedicated, caring individuals who assumed leadership roles at one time or another and who contributed greatly to the development of support services within the aviation industry. Those who served on, or are still serving on, aviation Critical Incident Stress Management teams should be justly proud of the gifts they have given their colleagues. As any beautiful building would be diminished if some of its bricks were missing, the Critical Incident Management field would be disorganized and ineffective if the collection of small, unnoticeable steps taken by others before us was absent from its background. This author thanks each one for their efforts and apologizes, in advance, for any oversights that may occur in this history.

Introduction

Recorded history is replete with descriptions of people helping people (National Bible Society 1994; Lee 2001). Some scholars argue in favor of a human condition of *genuine altruism*. They suggest that helping behaviors are an innate part of being human. In stark contrast, others argue that, if left on their own, human beings would rarely make efforts to help less fortunate people and that something besides innate human kindness would have to be the driving force behind their positive efforts. The most skeptical of scholars suggest that an incentive for some personal gain actually drives human selflessness. They note that, throughout the centuries, religious guidelines and their promises of rewards or punishment functioned as strong motivators for rendering help to others. The arguments will likely continue unabated since only those who assist others know what was in their hearts when they decided to take action.

No matter where one falls on the continuum of beliefs about helping, it is an undeniable fact that some people in all generations appear to be sensitive to the needs of others. They readily respond to distress and attempt to diminish it. It is also a commonly accepted belief that those who received help appreciate the efforts on their behalf.

Most people who help others feel personally fulfilled, but sometimes helping is costly. At the very least, assisting others takes time. It demands attention to the needs of others and absorbs physical and emotional energy. Indeed, some who helped others have sacrificed their own health and safety or even their lives to alleviate pain and suffering or to protect others from disease, disaster or death.

Even though every generation can point to examples of heroism, courage and self-sacrifice on behalf of others, it was not until the beginning of the last century that the approach to helping people in a state of crisis became organized and structured. *Crisis Intervention, psychological first aid, emotional first aid, critical incident stress management,* and *early psychological intervention* are a few of the terms currently used to describe the organized supportive assistance provided by trained people in times of acute distress (Neil, Oney, DiFonso, Thacker and Reichart 1974; Mitchell 2004). This chapter provides a brief history of the field of crisis intervention and explains how crisis intervention services entered into and influenced aviation.

The Roots of Early Psychological Intervention

Crisis intervention is an *active, temporary* and *supportive* process to assist people who are experiencing acute emotional distress. It is also known as *emotional first aid*

(American Psychiatric Association 1964; Neil, Oney, DiFonso, Thacker and Reichart 1974). The concept first entered psychological literature with Edwin Stierlin's article on a 1906 European mining disaster (Stierlin 1909). In the United States, in the same year, the National Save-a-Life League was founded in New York in an effort to prevent suicides (Roberts 2005). Lindemann (1944) and others expanded the concepts and applied it in warfare, disaster and situations involving acute grief (Salmon 1919; Lindemann 1944; Kardiner and Spiegel 1947; Artiss 1963; Parad and Parad 1968; Solomon and Benbenishty 1986).

The theoretical foundations and core principles of modern crisis intervention theory were developed by Gerald Caplan during the 1960s (Caplan 1961, 1964, 1969). Caplan suggested that *emotional first aid* or psychological first aid could be learned and applied by people who do not have formal training in psychology, social work or psychiatry. He believed that *excellent help in a crisis could come from paraprofessionals or from friends helping friends*, family members reaching out to other family members and communities supporting their own members. Caplan laid the groundwork for the supportive crisis intervention services that are currently provided by members of the clergy, by emergency services and by aviation personnel (Caplan 1961, 1964, 1969).

Crisis intervention services were predominantly provided on an individual, *one-on-one* basis through the mid 1970s. Group crisis intervention techniques were rare until the development of Critical Incident Stress Management (CISM) programs in the mid 1970s and 1980s. The utilization of groups was widely considered an innovative advance in the field of crisis intervention (Everly 1999). CISM programs incorporated both large and small group procedures as just a few of the many elements in an effective crisis intervention program (Everly 1999; Mitchell 2004; Roberts 2005; Gibson 2006).

Early Patterns of Aviation Psychological Services

Psychological services in the airline industry appeared to follow three general patterns established in non-airline organizations such as the military and police departments during and shortly after World War II. The first of the general patterns was that *psychological services first focused on screening of personnel*. Psychological testing of potential service members was important to assure the selection of the best people and to assist supervisors in assigning those personnel where they would provide the most effective and efficient work. Military and law enforcement organizations recognized that some of the work within their organizations was difficult and dangerous. Improper screening could contribute to a failed mission and the loss of lives. The military and police services wanted to make sure they did not have unstable people at the controls of deadly weapons or expensive equipment like ships, tanks and airplanes. The military, therefore, screened its personnel especially its pilots and law enforcement agencies screened potential police officers (McCarthy 2002; Reese 1987).

By the 1960s, the National Aeronautical and Space Administration (NASA) developed sophisticated psychological screening methods for the assessment of potential astronauts (Miller 2002). Soon the newly developing field of aviation psychology was mirroring both NASA and the psychological services of both the military and law enforcement services. Psychological screening became an integral part of the medical screening of most potential commercial pilots.

A second, but not well appreciated, pattern of *on-the-job evaluations* developed within the field of aviation psychology as it did in the military and law enforcement agencies (Reese 1987). This pattern included the use of aviation psychological services to protect the airline and the public and, although not stated, to enforce discipline and control. If a pilot's behavior was challenging to administration or if supervisors perceived it to be unusual or even remotely potentially dangerous, a pilot received a psychological evaluation. The results of the evaluation sometimes were used by aviation administrators to bar a pilot from flying or to fire the person altogether. For the most part, pilot psychological evaluations do indeed protect the flying public. Unfortunately, there have also been reports of abuse of the evaluation process.

A third pattern of psychological services emerged in the mid 1980s. It is *psychological support services* (Reese 1987). This pattern, by far, has enjoyed the most positive view from employees. It has greater acceptance than the initial job screening or on-the-job evaluation functions of aviation psychology. Crew resource-management programs improved cockpit communications and decreased safety risks. Psychological support services include alcohol and drug rehabilitation programs that allow pilots to fly again once they successfully complete a treatment program. Some airlines also developed family support programs for their employees (O'Flaherty 1995).

Critical Incident Stress Management in Aviation

Prior to 1990, and with the exception of the programs mentioned in the previous section, mental health professionals outside of the aviation industry predominantly expressed much of the interest they had in aviation by emphasizing the emotional needs of disaster survivors and the grieving loved ones of those who died (Duffy 1979; Freeman 1979). In the decades prior to 1990 peer-support crisis intervention teams were either unheard of or were just beginning their development within aviation.

Fledgling Employee Assistance Programs (EAP) were introduced in a few airlines in the early 1980s. By the end of the decade, most air carriers had Employee Assistance Programs. These programs usually consisted of mental health professionals who provided alcohol abuse assessment and intervention programs along with some general counseling. By this time, aircraft crew *resource management programs* were also working in some airlines. Those programs focused, however, on accident prevention, not on the management of the emotional needs of aviation employees.

There were certainly no standardized or comprehensive crisis intervention programs. In fact, the general atmosphere was generally negative toward airline employees who needed help with emotional distress. For example, in the United States, pilots who obtained psychological help for virtually any reason were sometimes considered emotionally unstable. In some cases, the Federal Aviation Administration took punitive action against stressed pilots and rescinded their license to fly. Some pilots suggest that this practice is still in effect today.

One notable exception to the description of the somewhat negative, or at least uninformed, atmosphere discussed above was the attention that was beginning to be paid to aircraft rescue and fire fighting crews and to law enforcement personnel working at air crash scenes (Forstenzer 1980; Mitchell 1982).

The Decade of the 1980s

In the early 1980s, a few airlines started to investigate crisis intervention programs and sometimes even sent key airline personnel for crisis intervention and stress management training. Such was the case in 1982 when Alaska Airlines and Wien Air Alaska sent pilots and human resource management personnel to a stress conference organized by the fire and emergency medical services organizations in Anchorage, Alaska.

In 1987, Michael LiBassi, then the manager of Employee Programs and Human Resources for American Airlines, invited the author of this chapter to provide an overview of crisis intervention and Critical Incident Stress Management to a group of approximately 25 American Airlines pilots, human resource personnel and managers at the American Airlines Headquarters at the Dallas-Fort Worth Airport. This was, perhaps, the first formal, aviation specific, crisis and stress management program on record (LiBassi 1987).

In 1988, Wien Air Alaska, Hawaiian Air Lines, Aloha Airlines and Alaska Air Lines personnel participated in CISM training at sites in Anchorage, Alaska; Seattle, Washington; Portland, Oregon, and Spokane, Washington. Selected staff from Air British Columbia received training in 1989. Greg Janelle, a flight attendant with Air BC developed a system wide support program for the airline in the following years. He is currently one of the most active aviation crisis intervention instructors.

Informal stress management programs for military aviation specialists were provided to the Norwegian Air Force as early as 1985. The very first military Critical Incident Stress Management team, however, was established in Rhine-Main US Air Force Base near Frankfurt, Germany in 1988. Other teams were soon established at Hickham Air Force Base in Hawaii and Wright Patterson Air Force Base in Ohio. It was not long before the US Army, US Navy, US Marine Corps and the US Coast Guard followed with programs of their own. Likewise, German Air Force and Navy personnel developed crisis support programs by the end of the decade.

Military aviation, however, entered the CISM picture in a significant manner just before the start of the Persian Gulf War in 1990. Certain wing commanders, chaplains and other operational support personnel such as psychiatrists, psychologists and family support personnel were concerned that there could be substantial loss of life in the Persian Gulf theatre. Many of them attended training programs provided by the International Critical Incident Stress Foundation (ICISF) in various parts of the United States. Military CISM training activity increased substantially about two months before the air aspects of the war started. Military aviation has remained active in military community and combat stress programs since the Persian Gulf War in 1990.

Military services in other countries are in the process of developing CISM teams within their services. For example, Singapore sponsored CISM training in that country in 2002. Chinese military officials attended a hospital sponsored CISM program in Hong Kong in 2003. The Portuguese Air Force established a CISM in 2005.

ALPA's Role in Crisis Program Development

The Air Line Pilots Association (ALPA) became involved in developing crisis support programs for its members in the early 1990s. Captains Alan Campbell, Robert Sumwalt and Mimi Tompkins researched and developed the original concept of a Critical Incident Response Program on behalf of ALPA. Dr. Donald Hudson, ALPA's Aeromedical Advisor, strongly supported the development of Critical Incident Response Programs (CIRP) within airlines. By May of 1994, ALPA's Executive Board unanimously passed a resolution formally recognizing and funding the program (Tompkins et al. 1996).

ALPA's endorsement of the International Critical Incident Stress Foundation's (ICISF) crisis training programs was a major step forward in the development of standardized comprehensive, integrated, systematic and multi-component crisis intervention and stress management programs for air carriers around the world. ALPA leadership contributed enormously to the expansion of crisis intervention programs within the airline industry. The progressive leadership in ALPA not only encouraged airlines to develop support programs for pilots, but for other aviation employees as well.

ALPA took a lead role in reducing the negative environment surrounding the issue of supporting stressed pilots. ALPA facilitated a significant flow of information to the National Transportation Safety Board (NTSB) and the Federal Aviation Administration (FAA) regarding the development of Critical Incident Response Programs for airlines. The psychological needs of pilots after critical incidents were often the focus of ALPA discussions with those agencies. ALPA invited certain officials from those governmental agencies to participate in Critical Incident Response Program (CIRP) training. The ALPA organization sponsored several specialized training programs for pilots and others associated with the aviation industry and encouraged its members to take the training from the International

Critical Incident Stress Foundation at its conferences around the United States and Canada (ALPA 1997).

Recent meetings between ALPA and the International Critical Incident Stress Foundation have clearly reaffirmed ALPA's enthusiastic support of CIRP in the aviation industry. In fact, as this book goes to print, an aviation-specific Critical Incident Stress Management (CISM) course is under development by a joint task force made up of ICISF, ALPA, the Airline Flight-attendants Association (AFA), The National Air Traffic Controllers Association, and Deutsche Flugsicherung (DFS) representatives (Howell 2006).

Flights Attendants Role in CIRP

Flight attendants make up a huge portion of the work force of most airlines. They have also been a driving force in the development of CIRP within the industry. Flight attendants who experienced traumatic events prior to the development of training programs and who had no support programs demanded that their airlines do something to help them (Purl 1986).

The Airline Flight-attendants Association (AFA), the largest organization representing the interests of flight attendants took up the issue and explored several options. They were most attracted to the ICISF training programs. Representatives from AFA attended ICISF trainings and began to apply CISM services within AFA and within several large airlines.

It is worth mentioning that when crisis intervention training became available flight attendants who suffered from long-term Post Traumatic Stress Disorder (PTSD) were often the very first to avail themselves of such programs. Many of them reported that the information alone was personally helpful. Information enabled flight attendants to put a name on the painful stress reactions they were suffering. It normalized the condition and helped many to employ effective personal stress reduction techniques (Purl 1996).

From the earliest training programs in the mid 1980s until current times, flight attendants from a wide range of airlines have been welcomed in CISM training programs sponsored by ICISF. The AFA in the United States has consistently supported training and peer support programs. Flight attendants from numerous airlines not represented by AFA have been key elements in the development of support programs for their own airlines (Bailey and Hightower 1996).

Most Major Air Carriers Have Crisis Teams

By 1995 American Airlines, Delta Air Lines, United Airlines, Northwest Airlines, US Air, Trans World Airlines and several smaller carriers had full peer-based crisis response teams. Southwest joined the growing group of airlines providing peer support programs in less than a year (Bailey and Hightower 1996). Even cargo carrying airlines such as FedEx had well-organized crisis support programs in place by the end of the decade. The FedEx program description states:

> Nearly all U.S. airlines, both passenger and cargo, are now developing standardized CIR programs.... (Dillenbeck 1996, p.1).

Air carriers in countries around the globe have also developed crisis response teams. Qantas Airlines, for example, developed a peer support program in its Pacific operations around 1992. Mental health professionals within the airline worked closely with the trained peer support personnel. Employee satisfaction with the program was high.

Lufthansa was one of the first to develop a European team and the program today is one of the most active and effective in Europe. Lufthansa is committed to fostering crisis intervention programs in other airlines. It demonstrates its commitment by assisting other airlines after distressing events and by its active participation in programs such as the international aviation conference sponsored by the University of Graz, Austria, in 2004.

The orientation and education of many CIR programs in nations outside of the United States was enhanced by yearly training programs sponsored by Delta Airlines at its training academy in Atlanta, Georgia. The programs started in 1994 and were under the direction of Tara Woolridge. ICISF instructors accepted invitations to the academy to present crisis intervention programs and airlines from around the globe selected personnel to attend the training in critical incident management. Airlines from Austria, Canada, Czech Republic, England, Finland, Germany, Ireland, Italy, Norway, Poland, Singapore, Sweden, the United Kingdom and the United Arab Emirates were among the many airlines that participated in the training programs. Pilots, flight attendants, mechanics, ticket agents, managers, safety engineers and many others benefited from the educational programs. A small stream of interest had turned into a flowing river within a decade. Positive reports from aviation industry personnel have been consistent over the nearly 20 year history of these crisis programs.

Trial by Fire

Interest in providing support services to survivors or the loved ones of deceased air disaster victims existed in several large airlines such as Pan American and Trans World Airlines before the development of comprehensive staff support programs within the aviation industry. For many other airlines, however, organized efforts to assist these primary and secondary victims did not occur until 1992. In that year the American Red Cross initiated efforts to create a national disaster mental health network in the United States and signed agreements with the American Psychological Association (APA) which provided assistance in developing professional guidelines for psychological assistance in disasters. The APA continues to serve as a source of mental health professionals who assist the Red Cross in the aftermath of natural and technological disasters. Later, the American Red Cross would develop specialized training specifically for managing the psychological aftermath of aviation disasters.

In the late 1980s Johanna O'Flaherty, who was thinking of trauma intervention for the loved ones of aviation disaster victims long before the concept became popular for most airlines, developed Pan American Airlines' Trauma Response Team. The team was made up of airline employees who volunteered to assist families in the event of a tragedy. The team was system wide and included pilots, flight attendants, ticket agents, managers and other airline employees. The Trauma Response Team was built upon Dr. Gerald Caplan's philosophy that excellent crisis intervention services could be provided by paraprofessionals (Caplan 1961, 1964, 1969). The team was pressed into service in the aftermath of the horrific Lockerbie air disaster in 1988. Unfortunately, and despite the success of the important services rendered to suffering family members, Pan American Airlines did not survive the political and economic fallout of the crash.

Johanna O'Flaherty then took a similar position with Trans World Airlines. She set to work to establish a Trauma Response Team within that airline (O'Flaherty 1995). In October of 1995, O'Flaherty invited The International Critical Incident Stress Foundation to present Critical Incident Stress Management (CISM) training to the existing TWA Trauma Response Team. Her plan was for the Trauma Response Team to have two branches. One would provide emotional first aid to the family members of deceased or injured passengers. The other branch of the program would provide crisis intervention services to employees of the airline. No one could have predicted at that time how quickly the team would undergo a trial by fire. To the shock of the airline, and the world, TWA flight 800 was lost just nine months later in July of 1996.

The Trauma Response Team immediately mobilized and flew to New York to set up a Family Assistance Center. TWA employees, despite the overwhelming losses of 54 of their colleagues, worked tirelessly – and at times nearly 20 hours a day – to provide information, crisis support, guidance, transportation, operational briefings and a host of other services to the shocked and grieving family members.

O'Flaherty recognized that her team was stressed and saddened by their losses and were growing fatigued by the long hours and the overwhelming volume of work facing them. She issued a call for assistance from the crisis management teams of other airlines. A dozen or more airlines sent personnel to work side by side with the TWA staff. No one requested remuneration; no one engaged in competition. Every airline donated its personnel to the crisis management efforts for the duration of the acute phases of the disaster. Airlines were not the only organizations to respond to O'Flaherty's call for help. The Fire Department of New York assigned Lt. Douglas J. Mitchell as an on-site Critical Incident Stress Management coordinator to assist O'Flaherty in running the operation. ICISF sent several mental health professionals to advise and assist with the effort. The many families served by the Trauma Response Team were better able to bear the turmoil and grief associated with the disaster because of the dedicated care provided by TWA's Trauma Response Team. TWA employees who were served by the combined efforts of more than a dozen

Critical Incident Response Programs later expressed their gratitude for the services and praised the efforts of the combined airline crisis teams.

The 1996 TWA 800 crash off the coast of Long Island, New York, clearly demonstrated the need for crisis intervention services for victims and their families. In a governmental review, the Gore Commission Report on aviation safety and security also recommended a broad spectrum of crisis intervention programs to assist the families of aviation disaster victims (Gore 1997).

Air Traffic Controllers and CISM

Tragedy, again, was the backdrop for the development of Critical Incident Stress Management programs for air traffic controllers and others. Transport Canada initiated an air traffic controller Critical Incident Stress Management program around 1987. At first, the program provided individual crisis intervention services to distressed air traffic control throughout the Canadian air traffic control system. The program was well-received by the employees who requested additional training and an expansion of the program. In 1989, United Airlines experienced the Sioux City, Iowa crash. Because airline Crisis intervention teams had not yet been developed, fire, emergency medical and law enforcement CISM teams assisted both the crash-rescue personnel in Sioux City and airline employees in 13 locations around the United States in managing the psychological impact of the event.

Canadian air traffic controllers mobilized and crossed the border into the United States. They provided well-organized and effective peer support services to their American counterparts in Sioux City. It was clear that the Canadian team members did not perceive national borders as a barrier to helping out their colleagues after they experienced a disturbing air crash (Dooling 2003).

It is a little known fact that the Sioux City disaster had far-reaching effects on the field of Critical Incident Stress Management both within and outside of air traffic control systems. At the time, CISM teams were a loose knit network of newly formed teams in several countries, predominantly the United States and Canada. These teams worked independently of each other. They rarely met to exchange ideas and few of the teams had more than two or three years experience in crisis intervention and traumatic stress management. Some had more or, sometimes, considerably less training than other teams and there were serious gaps in the standardization of their training. Despite the lack of organization, 13 CISM teams from ten states responded to telephone calls from this author requesting that they assist United Airlines personnel in various flight centers around the United States.

Within four months of the Sioux City crash, the American Critical Incident Stress Foundation (renamed the International Critical Incident Stress Foundation in 1992) was established. Today the primary purposes of this organization remain education, training, coordination, crisis intervention services, networking and standardization of CISM teams throughout the world.

The Sioux City disaster was among several factors that stimulated the development of the US Federal Aviation Administration's air traffic control CISM team under the direction of Brad Troy. The CISM team is unique and serves as a model program in that the National Air Traffic Controllers Association, which is the air traffic controllers union, and the Federal Aviation Administration (management) jointly sponsor it. In fact, both parts of the organization worked together to develop an informative 20 minute video taped program on Critical Incident Stress Management for air traffic controllers (NATCA/FAA 2003).

The team is currently comprised of mental health professionals from the FAA's Employee Assistance Program and air traffic controllers who volunteered for the program. The team received CISM training in the early 1990s and responded to numerous incidents including the devastating terrorist attacks of 11 September 2001.

EUROCONTROL, the European Organization for the Safety of Air Navigation has a keen interest in the development of European air traffic control crisis intervention services and standards for CISM programs in ATC. For approximately a decade, it has sponsored several important studies on the topic as well as crisis intervention and staff support presentation at conferences. It encourages air traffic control systems to develop support programs for their personnel.

German air traffic controllers were the next professional aviation organization to benefit from the services of a CISM team. The German Air Navigation Service Provider – Deutsche Flugsicherung (DFS) – team, directed by Jörg Leonhardt, was developed in 1998. Jörg Leonhardt, who is involved in safety management for the German air traffic control system, is one of the author-editors of this book. He is a well qualified instructor who is approved by ICISF to provide CISM training. He has worked with many European air navigation service providers (ANSPs), supported them in developing a CISM program and trained peers from different nations. The DFS team today has 80 peer support personnel and is very well educated in crisis intervention. Every member of the DFS team holds multiple certificates of training from ICISF. The members of the team have had CISM training in individual support services as well as two levels of group crisis intervention. It is one of the most well trained teams in aviation. German air traffic controllers have reported very high levels of satisfaction with the services of the team members.

The team has responded to many smaller events, but also experienced a trial by fire in two major incidents. One was an airplane crash into terrain and the other was the collision of two aircraft over Lake Constance in Switzerland in 2002. Again, national borders became a minor secondary consideration when compared to the emergency call to assist colleagues after a terrible tragedy.

In cooperation with the International Federation of Air Traffic Controller Association IFATCA DFS has organized an international ATC conference in Bucharest, Romania in 2004. At this conference the experience from different ANSPs were presented to other Air Traffic Controllers. As a consequence some ANSPs started their CISM program following the Bucharest conference. Again the controllers associations as well as the Pilot association and the association of the flight attendants play a vital role in introducing CISM to their professions.

DFS has also delved into research in the field of CISM. To date, DFS has collaborated with the University of Dortmund and the University of Copenhagen on two studies regarding the economic advantages of providing peer support. One of those studies, which demonstrated that the economic benefits of a well run air traffic control peer support program surpassed the costs of developing the program, was published in the *International Journal of Emergency Mental Health* (Vogt et al. 2004).

In a rather unusual program, Frankfurt International Airport and DFS work together in a mutual aid arrangement. They share CISM training opportunities, meet together regularly and support each others efforts. For the last six years before this publication DFS and Frankfurt Airport have collaborated on a yearly international crisis intervention conference for the aviation industry. There are plans to continue the program. In the unfortunate event of a major emergency both the Frankfurt Airport and DFS will share resources to provide CISM services to employees and the travelling public. The crisis intervention team is approved by the ICISF and called *ATC-AP* (Air Traffic Control – Airport).

Finally, DFS has been instrumental in the opening of an office of the International Critical Incident Stress Foundation in Europe. The office will be opening officially in the autumn of 2006. DFS is providing office space and some staffing to help get the office established. The opening of the ICISF office in Europe constitutes a historical benchmark in the development of CISM teams throughout the world. It will be the first time an official ICISF office will be open in any country in the world. DFS's support of the office will enhance communications with ICISF headquarters in the USA.

NAV Portugal has a well-organized and trained CISM team. The team members have demonstrated their crisis skills in numerous small-scale events. The team members have gained the respect and appreciation of their colleagues throughout the Portuguese air traffic control system.

Frankfurt Airport

This chapter has discussed airline CISM teams, the role of ALPA, and the activities of air traffic control CISM teams in Canada, the US and in Europe. Before concluding, it is important to recognize the fact that airports play a vital role in CISM services. They are the base of operations for airlines and most of the services associated with aviation. Airports should not be left out of the formula for successful CISM teams.

The Frankfurt International Airport in Germany developed a CISM team to serve all airport employees. As mentioned earlier in this chapter, the airport works closely with the German air traffic controllers to provide crisis support services in the event of a major air incident. The team has members who have been carefully trained in

CISM. They coordinate their efforts with the medical facility at the airport and are on call for any situation that demands psychological first aid.

Conclusion

From the airport facility to the airlines and from the ground staff to the pilots, air traffic controllers, and flight attendants, the aviation industry has proven that *excellent help in a crisis could come from paraprofessionals or from friends helping friends*. Whether it is a situation involving personal grief, a family crisis or an air disaster, aviation CISM team members respond with skill, care and concern. They have demonstrated the enormous value of crisis intervention teams and the number of programs is likely to expand in the coming years. They make a difference in the lives of others.

References

Air Line Pilots Association (ALPA) (1997), *Critical Incident Response Program Guide*. (Herndon, VA: ALPA. H).

American Psychiatric Association (1964), *First Aid for Psychological Reactions in Disasters*. (Washington, DC: American Psychiatric Association).

Artiss, K. (1963), Human behavior under stress: From combat to social psychiatry. *Military Medicine* 128, 1011-1015.

Bailey, W. and Hightower, S. (1996), *Critical Incident Response Program*. (Dallas, Texas: Southwest Airlines Pilots' Association, Transport Workers Union).

Caplan, G. (1961), *An Approach to Community Mental Health*. (New York: Grune and Stratton).

Caplan, G. (1964), *Principles of Preventive Psychiatry*. (New York: Basic Books).

Caplan, G. (1969), Opportunities for school psychologists in the primary prevention of mental health disorders in children, In A. Bindman and A. Spiegel (Eds.) *Perspectives in Community Mental Health* (pp.420-436). (Chicago: Aldine).

Dillenbeck, P. (1996), *Critical Incident Response Program for FedEx Flight Crewmembers*. (Memphis, TN: Federal Express Corporation).

Dooling, M. (2003), *Critical Incident Stress Management* (A videotaped presentation at EUROCONTROL's conference on "Training of controllers in Unusual / Emergency Situations, 1996). Luxembourg: EUROCONTROL, Institute of Air Navigation Services.

Duffy, J. (1979), The role of CMHCs in airport disasters. *Technical Assistance Center Report* 2:1, 1; 7-9.

Everly, G.S., Jr. (1999), Emergency Mental Health: An Overview. *International Journal of Emergency Mental Health* 1, 3-7.

Freeman, K. (1979), CMHC responses to the Chicago and San Diego airplane disasters. *Technical Assistance Center Report* 2:1, 10-12.

Forstenzer, A. (1980), Stress, the psychological scarring of air crash rescue personnel.

Firehouse July, 50-52; 62.

Gibson, M. (2006), *Order From Chaos: Responding to Traumatic Events*. (Bristol: UK, The Policy Press, University of Bristol).

Gore, A. (1997), *White House Commission Report on Aviation Safety and Security*. (Washington, DC: The White House).

Howell, D. (2006), *President, International Critical Incident Stress Foundation* (ICISF). Personal Communication.

Kardiner, A. and Spiegel, H. (1947), *War, Stress, and Neurotic Illness*. (New York: Hoeber).

Lee, S.C. (2001), *Backup On The Beat: An Inspiring Collection of Stories, Essays & Thoughts for America's Peace Officers*. (Rocklin, CA: Pascoe Publishing).

LiBassi, M.D. (1987), *Employee Trauma Recovery Program. Personnel Resources, Local Procedures Manual*. (Dallas, Texas: American Airlines).

Lindemann, E. (1944), Symptomatology and management of acute grief. American *Journal of Psychiatry* 101, 141-148.

McCarthy, P. (2002), Dr. Patrick McCarthy's brief Outline of the History of I/O Psychology. http://www.mtsu.edu/~pmcarth/io_hist.htm.

Miller, K. (2002), *How Astronauts Get Along*. (Houston, Texas: Science@NASA.gov).

Mitchell, J.T. (1982), The psychological impact of the Air Florida 90 Disaster on fire-rescue, paramedic and police officer personnel. In R Adams Cowley, Sol Edelstein and Martin Silverstein (Eds.) *Mass Casualties: a lessons learned approach, accidents, civil disorders, natural disaster, terrorism*. (Washington, DC: Department of Transportation (DOT HS 806302), October, 1982).

Mitchell, J.T. (2004), Characteristics of Successful Early Intervention Programs. *International Journal of Emergency Mental Health* 6, 4. 175-184.

National Bible Society, (1994), *God's Word for Peace Officers*. (Grand Rapids, MI: World Publishing).

NATCA/FAA (2003), *You are Not Alone: Critical Incident Stress Management Program*. (Washington, DC: NATCA / FAA).

Neil, T., Oney, J., DiFonso, L., Thacker, B. and Reichart, W. (1974), *Emotional First Aid*. (Louisville: Kemper-Behavioral Science Associates).

O'Flaherty, J. (1995), *Trans World Airlines: Trauma Response Team, Section Leader Manual*. (St Louis, MO: Trans World Airlines, Inc.).

Parad, L. and Parad, H. (1968), A study of crisis oriented planned short-term treatment: Part II. *Social Casework* 49, 418-426.

Purl, S. (1986), *Am I Alive?: A Surviving Flight Attendant's Struggle and Inspiring Triumph Over Tragedy*. (San Francisco, CA: Harper & Row, Publishers).

Purl, S. (1996), *A flight attendant air crash survivor and author of Am I Alive?* Personal communication.

Reese, J.T. (1987), *A history of police psychological services*. (Washington, DC: US Department of Justice, Federal Bureau of Investigation).

Roberts, A. (2005), Bridging the Past and Present to the Future of Crisis Intervention and Crisis Management. In Allen Roberts (Ed.) *Crisis Intervention Handbook:*

Assessment, Treatment and Research. Third Edition. (New York: Oxford University Press).

Salmon, T.S. (1919), War neuroses and their lesson. *New York Medical Journal* 108, 993-994.

Solomon, Z. and Benbenishty, R. (1986), The role of proximity, immediacy, and expectancy in frontline treatment of combat stress reaction among Israelis in the Lebanon War. *American Journal of Psychiatry* 143, 613-617.

Stierlin, E. (1909), *Psycho-neuropathology as a result of a Mining Disaster March 10, 1906.* (Zurich: University of Zurich).

Tompkins, M., Morin, W., Campbell, A., Hudson, D., Sumwalt, R., Johnson, J., Anding, J.D. and Bracken, J. (1996), *Air Line Pilots Association: Critical Incident Response Program Manual.* (Herndon, VA: ALPA).

Vogt, J., Leonhardt, J., Köper, B. and Pennig, S. (2004), Economic Evaluation of CISM – A Pilot Study. *International Journal of Emergency Mental Health* 6:4, 185-196.

Chapter 7

Critical Incident Stress Management in Air Traffic Control (ATC)

Jörg Leonhardt

This chapter provides an overview of the implementation process of CISM in the Air Navigation Services; it describes the special requirements for air traffic control in this field and explains the advantages of a peer-based program.

CISM in ATC

Within Air Traffic Management (ATM) and Air Traffic Control (ATC), CISM has gained importance since the Sioux City accident in July 1989. After the crash, the Air Navigation Service Provider of Canada (NAV Canada), which had established a CISM program for their Air Traffic Controllers in 1987, crossed the border and supported its American colleagues in Sioux City (see Chapter 6).

Thanks to the dedicated work of Mike Dooling, NAV Canada was one of the first air navigation services organizations to establish a CISM program. Since then, several Air Navigation Service Providers (ANSPs) have followed the Canadian example and have started implementing their own CISM programs. Some ANSPs realized the necessity of an established CISM program without having experienced a major accident, while others started with the implementation after the organization had dealt with a major accident. The German ANSP (DFS Deutsche Flugsicherung GmbH) took the opportunity to learn from the experience of other ANSPs and therefore decided to implement a CISM program without having been involved in a serious or disastrous accident.

DFS started to plan and implement a CISM program for the operative personnel in 1997, mainly for the Air Traffic Controllers (ATCOs), but also for the flight data assistants and technicians. Based on the publications and recommendations of the European Organization for the Safety of Air Navigation (EUROCONTROL) (Woldring and Amat 1997), DFS established the CISM program according to the standards of the International Critical Incident Stress Foundation (ICISF). Since 2000, DFS can rely on more than 80 trained and ICISF certified peers, who offer their colleagues CISM measures in all operational units 24 hours every day.

Special Requirements for ATC

Air traffic controllers (ATCOs) work in a high-reliability business. This means maintaining constant vigilance, high concentration, stress resistance and also team performance. Besides the mental and team-related requirements for ATCOs that demand enormous energy, critical incidents always depict extraordinary situations that go beyond the everyday stress and the resulting strain.

Due to their education, training and a healthy stress management, ATCOs are able to deal with the special demands and the resulting strain. An ergonomic work organization with, for example, forward-rotating shift systems, and breaks in line with their position's workload can help them to cope with the daily work stress (Vogt and Kastner 2001). These issues are continuously communicated to and discussed with the ATCOs as early as at the beginning of the training and, therefore, professional ATCOs are very familiar with them. ATCOs actively manage their own stress in terms of an individual maintenance of performance and well-being.

Nevertheless, some incidents and the related sensations and reactions might exceed the individual stress tolerance and coping capacity. A comprehensive CISM program is required to help ATCOs deal with these Critical Incidents, which cannot or can only be barely dealt with by ATCOs on their own.

The debate on stress and stress reactions especially in this particular field of work has lead to an awareness of critical incident stress reactions. Therefore, the possible reactions after a critical incident are familiar to the ATCOs. They are informed about the different types of critical incident stress reactions and the consequences – both of a personal and professional nature. During the information campaign, as the starting point of the CISM implementation plan, all ATCOs are informed in briefings, in documents and during presentations at the local operational units. Dealing with critical incident stress reactions is now part of the professional self-image of ATCOs in Germany, as will be the case in every country where CISM becomes an established program.

The requirements on a CISM program in air traffic control (ATC) differ in three major aspects compared to other occupational groups:

- Critical incidents in ATC do not necessarily mean an accident has actually happened.
- Generally, the ATCO working in the Area Control Center (ACC) and Upper Area Control (UAC) is separated from the incident and has no direct sensory impressions. Only the tower controller may witness an accident on the runway with his or her own eyes.
- Due to their jobs, ATCOs are directly involved in the development and progress of a Critical Incident and usually feel responsible.

The three aspects will be further explored in the following sections.

Critical Incidents in ATC are not Necessarily an Actual Accident

In the opinion of an ATCO, a Critical Incident is, for example, a separation loss between two aircraft. As a rule, the safety margin is big enough to ensure that falling short of the separation minimum does not lead to an accident. Normally, it is not even noticed by the passengers or the public. However, an unintended separation loss, even if it is only marginal, can be a major threat to the self-image of an ATCO and can lead to Critical Incident Stress Reactions.

The ATCO's *inner eye* is viewing a catastrophe and the fantasy completes the inner picture. This is a mental picture and does not necessarily correspond to reality. But it is the nature of our brain trying to complete a picture. In training and professional experience the ATCO develops pattern recognition for potentially critical situations. Due to this early warning system, the ATCO is aware of the potential danger an unintended separation loss may have if there are other aircraft around and the situation becomes increasingly critical. Even if there has been no direct danger of a crash, the ATCO's mental early warning system has already raised the alarm. The anticipation of what might have happened can trigger the inner pictures, which are truly experienced and can therefore lead to Stress Reactions. The limbic system (see Chapter 4) triggers fight or flight reactions of the body. The imagination activates physiological reactions of the brain and cause Critical Incident Stress Reactions. Although the situation caused no accident or harm to others, the stress reactions are real and potentially harmful to the ATCO.

The incident may be barely perceived (for example by pilots) or not perceived at all (for example by passengers, public). The ATCO's subjective experience of the incident and the adherent stress reactions do not correspond to the reality as perceived by others. Thus, ATCOs talking to people outside ATC about an incident they have experienced as critical may be confronted with lack of judgment, misunderstanding and ignorance. In such a situation, potential helpers who are not familiar with the special requirements of psychological first aid to ATCOs – even if they are professional physicians, psychologists or spiritual advisers – may try to play down the incident by relating it to real accidents.

Often people face their own inability to help and try to comfort the affected people in trying to convince them that nothing happened and there is no reason for such reactions (see Chapter 5). As a consequence, they would make the situation even worse by stimulating the mind to imagine what could have happened. ATCOs will feel even more misunderstood during such a discussion and try to avoid talking about their actual reactions and symptoms. The ATCOs are often very confused and believe that something is wrong with them and their perception of reality.

Naturally, this might lead to enormous consequences as an offensive debate and coping mechanisms are missing. The ATCOs might withdraw and hide their true feelings and reactions. The feeling of "I am not normal" or "Something is wrong with me" becomes overwhelming and intensifies the crisis. If nobody else regards the incident as problematic and if it seems there is no reason to worry or go crazy, the ATCOs' own reactions are increasingly perceived as abnormal. Self-medication (for example with alcohol, sleeping pills, and sedatives) aggravates the situation and a vicious circle begins: The symptoms stay or deteriorate; medication is increased and so forth. Some ATCOs try to hide the symptoms to the public and the self-treatment continues. The longer this vicious circle persists, the harder it is to ask for help and exit the circle. The ATCO's belief of being abnormal and self-doubts get out of control and it becomes increasingly difficult to put the symptoms into perspective with the incident.

The symptoms might become chronic (see Chapters 3 and 4) – in the worst case a Post Traumatic Stress Disorder (PTSD) – will develop, and the original relation between symptom and incident becomes blurred. In such a worst case, it is common to only treat the physical symptoms with medication or surgery. In such a state, it is difficult to relate these disorders to an incident which may have happened years ago.

In the case of an accident noticed by the public, this negative spiral is less likely to develop as the individual reactions correspond to the perceived reality. The reactions of the people affected are assessed as *normal*.

Besides the reactions perceived to be unusual, and which cause some self-doubts, special attention has to be paid to the blow to the professional self-image of ATCOs, particularly as ATCOs have experienced similar situations before without strong stress reactions. Traditionally, it has been the culture of air traffic control to view critical incidents as part of the job which can be coped with alone. Modern ATC needs a shift from this *lonely rider* or *cowboy mentality* to a team-based culture, in which learning also includes learning from failure, which is organized and not sanctioned.

Self-doubts have an intensifying effect: "I have dealt with such situations before", "I am not sure anymore if I am capable of doing this job", "I have experienced worse than this and managed just fine".

Self-doubts intensify the feeling of *not being normal*, as others deal better with the situation or the person affected used to deal better before. Self-doubts depict an obstacle to seeking professional help.

The ATCO Working in ACC and UAC is Physically Separated from the Incident

Critical situations usually do not lead to an accident. If the unlikely actually does happen, ATCOs working with radar are excluded from all direct sensations. Smell, sound, taste and visual sensations are not perceived. After an accident has happened on the radar screen, the ATCO imagines pictures and sensations of the events happening on the site. Therefore, it is not possible to talk about direct sensations during a CISM intervention. However, the pictures the ATCO has imagined and the involved (imagined) sensations and perceptions have to be addressed.

The visualizations do not necessarily correspond to the actual events and might lead to a deferral of perception in the long run. The media partly contribute to an adjustment of the inner pictures and visualizations; nevertheless, the media show a subjective reality as well and do not always provide objective reports. However, in terms of a CISM measure, it is crucial to support the ATCO in the adjustment of the subjective reality.

In January 2004, a passenger aircraft approaching Munich Airport made an emergency landing. The forced landing took place on an open, snowy field, controlled by approach control and the tower controllers at Munich Control Tower, but not in the view of the controllers. After it was announced that the aircraft had landed according to emergency procedures and that nobody had been injured or died, managers of Munich tower called for CISM measures. Besides the applied measures (for example SAFER-R; see also Chapters 3 and 5), ATCOs could visit the landing site to take a look at the respective aircraft, speak to the pilots and so on. The benefits of this procedure will be described in the following section.

The ATCOs are Involved in Critical Incidents Automatically Due to their Work

Taking a look at the aircraft, getting a picture of the real extent of the damage, and being able to adjust their own imaginations by seeing the condition of the passengers and speaking to the pilots greatly contributed to stabilizing the ATCOs. Comparing one's own imagination with reality, with the inherent reflection with the incident and the stabilization, is the first step towards coping and reducing the effects of the stress reactions.

The pilots gave the ATCOs positive feedback on their professional support and work. For this reason, the ATCOs were able to experience their work as an important contribution instead of a failure. Thus, the ATCOs obtained a holistic view of the situation and their own contribution. Realizing that their professional work gave everybody a lucky escape helped the ATCOs to recover from the critical incident stress reactions.

Also in cases of tragic accidents with death and/or injury, it is important to compare the inner pictures and to be aware of one's own contributions. Feelings of guilt play an important role in this context. It has to be assumed that ATCOs always make high demands on air safety and act accordingly. Feelings of guilt are understandable as they highlight the ATCOs' responsibility for their work. However, guilt in a legal sense is another story than feelings of guilt, and it is not part of a CISM intervention to inquire or to investigate guilt.

When and to what extent reality is closely analyzed depends on the situation, the reaction of the ATCO, the expected consequences, and the peers' ability to cope with stress. Comparing one's own perceptions with reality does not mean confronting oneself directly with the worst. Instead, it is a step-by-step approach to the incident, the accident, and the negative consequences. This approach may take several weeks or months and should happen with professional support (psychologist, clergyman, physician, or therapist) and does not belong to the tasks of the peers. However, the peers should make sure that the ATCO receives professional support if necessary – this is called referral in the CISM terminology. Facing the reality is an important part of a lifelong recovery from critical incident stress and an important contribution to the prevention of PTSD.

The professional involvement of the ATCO in the incident usually causes self-doubts: "Have I done everything correctly?" "Shouldn't I have tried this or that?"

With the benefit of hindsight, it is much easier to rethink the decisions because the results and how they came about are known. At the time, however, decisions must be made with little time and information. Everyone in this field creates safety through practice and striving to do a good job. This is especially true for air traffic control and aviation which is highly dynamic and also complex as many different and interacting factors have to be considered. The ATCOs are part of this system and significantly contribute to the safety of the system. However, they are also subject to the contextual factors and can only partly influence them.

Nevertheless, feelings of guilt play an important role for the ATCOs. Dealing with the feelings of guilt, overcoming them and adjusting to reality are fundamental elements of CISM intervention. Peers are able to understand the feelings of guilt, the reactions after a critical situation and the impact on the professional self-image of such an incident. Peers can therefore normalize such reactions and gain the necessary trust from their colleagues in terms of these abilities (see Chapter 5). These elements have to be one of the main points of focus in the training of the peers (see Chapter 11).

After having explored the three requirements on a CISM program in ATC, we now turn to the implementation process of CISM. This includes the identification of suitable peers, their training, the structure of the program and the related experience.

Implementation

Before a CISM program is implemented, an initial and conception phase take place. During the initial phase, the CISM manager names people who will be responsible for the implementation and the establishment of the program. At DFS, it is standard procedure for the management to initialize the first phase in order to add importance and support for the program (see Chapter 2). The appointed person in charge will become familiar with the methodology and gain support from experienced colleagues.

During the conception phase, relevant steps and their implementation are defined. The plan should comprise the following aspects:

- Embedding of CISM into the culture of the ANSP
- Methodology and processes of CISM
- Definitions and basic understanding of CISM, CI, CIS
- Model of the peers and election/selection of the peers
- Training and education of the peers
- Certification of the peers
- Training of the supervisors and managers
- Standards and procedures

- Crisis Intervention Team
- Information campaign within the organization
- Quality management.

Selection/Election of Peers

Identifying suitable peers can take two paths: selection or election. For the selection, candidates apply for the job. They go through different selection procedures and are trained afterwards. For the election, colleagues and/or managers nominate their candidates, or those interested nominate themselves and are voted in by the staff. After their election, they start the CISM training. Both procedures have been applied in different ANSPs and both proved successful. However, each procedure has certain advantages over the other.

The election procedure focuses on the trust in the candidates. Is the candidate the one who is believed to be capable of fulfilling the role as a peer, is this person the one to turn to in case of an incident or an accident? Can the elected candidate be trusted in terms of treating personal feelings and statements confidentially?

The election procedure should be applied if CISM is to be established within an organization in which a less constructive culture of dealing with failures can be found. In an organization having difficulties in dealing with feelings of the employees, the election procedure is the appropriate solution. During the training, the focus is on the skills of the peers and the CISM methods. There is no later assessment or selection of the people who finished and completed the peer training.

In the selection procedure, choosing *appropriate* peers is the main point of focus: They should be emotionally stable, capable of dealing with the emotions of others and live up to the requirements of a peer. For the selection procedure, it is very important to have experienced CISM Mental Health Professionals, who are familiar with ATC, as members of the selection team. A solely psychological selection with the common procedures does not cover the whole range of selection criteria for the peers. After the selection procedure, the training should focus on information and self-promotion. It is essential to build up trust in the peers and their role.

The selection of the peers has to be well considered and should center on the culture of communication and management of failure within the ANSPs.

Training and Education of the Peers

The peers have to be well trained. They are employees filling an additional role for which they have to be trained. The training should be carried out according to the established and proven standards of the International Critical Incident Stress Foundation (ICISF) and be conducted by an ICISF certified trainer. The training and education take place over three to four years based on the following courses:

- Individual Crisis Intervention and Peer Support

- Basic Group Crisis Intervention
- Advanced Group Crisis Intervention.

There should be a 12 to 15 month break between courses in order to apply the methods learned. Refresher courses and peer supervision meetings are optional, but highly recommended. The course contents are listed in Chapter 11. The courses are completed with an internationally recognized certificate of the ICISF. The certification of the ICISF is important for the organization as it ensures an established and internationally recognized training of the peers. The certification is important for the peers as it gives the peers confidence and trust in their work. Established and high-quality training represents the quality of the organization and its CISM program. This aspect is not to be neglected (see Chapter 2).

The Peer Model

The following section describes the peer model in CISM. It is based on another Ashgate book (Leonhardt 2005).

Critical Incident Stress Management centers on the involvement of peers from the same professional environment. The peer model offers several key advantages:

- Colleagues are often accepted more quickly than individuals from outside the profession such as psychologists, physicians, members of the clergy and pastoral workers.
- Peers from the same occupational group appreciate the individual's CIS reactions better because they may understand them better – the peer may even have experienced similar reactions.
- Peers have a better understanding of the facts and circumstances of the event because they have the same professional qualification and socialization. This is especially true with respect to the above mentioned discrepancy between public opinion and ATCO view on critical incidents.
- The understanding of feelings of guilt is easier for peers.
- Peers are more able to normalize the stress reactions and minimize the unwanted feelings because they come from the same background and have had similar experiences.
- Peers speak the same professional *language*.
- An action to address the situation is often more credible and more easily accepted if proposed by a peer.
- The involvement of health professionals (for example physicians, psychologists) may give the impression that the individual is *ill*, which is not the case with peers.
- Peers can often be called in more quickly and without complicated or bureaucratic procedures.

Peers are not mental health professionals, therefore their abilities are limited:

- The identification with the colleague and the critical incident may trigger critical incident stress reactions amongst peers. In the worst case, the peer's professional aptitude may be impaired.
- The fact-finding phase might become overly comprehensive, with both parties becoming too involved in discussions of detail and procedure.
- Emotional phases in CISM, namely the reaction phase, may be too brief or even skipped entirely, since both parties may consider them as *threatening*.
- People or procedures may be left unduly exposed to criticism.
- Too little consideration is given to self-motivation as a peer.
- The limits of the discussion are not respected – often it may be difficult to avoid straying into personal matters.

However, the advantages clearly outweigh the disadvantages and the latter can often be limited or even eliminated through good peer management. Peer supervision and peer debriefings are necessary to help peers in overcoming their identification with the reactions of the colleagues as well as to give them the opportunity to have a break and to recover.

Structure

Parallel to selection/election and training of peers, a clear structure for the application of CISM interventions and the respective rules and procedures needs to be established. It can be assumed – as will be later described in Chapter 12 – that the critical incident stress reactions of an ATCO alter his ATC skills. This does not necessarily mean that the ATCO is unable to work, but it implies that any incident can become critical in the sense that it impairs the ATCO's well-being and performance potential. Therefore, a realistic self-assessment of the ATCO with respect to work ability as well as the leadership skills of the watch supervisors and the professional behavior of the peer is paramount. Watch supervisors and peers must be able to assess the actual stress coping capacity of the ATCO bearing in mind the social and family circumstances. But the question arises as to how can the self-assessment of the ATCO be corrected if needed?

This can be achieved by clear procedures and distinctive rules within the CISM program. The rules and procedures are designed to ensure that the assessment of one's own ability to perform, the consultation of a peer or manager and the decision to adopt a CISM measure are understood as normal procedures.

If everybody sees the necessity of CISM, if everybody feels that people affected as well as their colleagues and the organization as a whole profit from CISM, it will be easier to acknowledge CISM interventions as a standard procedure. Not every exceptional situation is a critical incident, but every situation can turn into one, even if it is not exceptional.

As the consequences of critical incident stress reactions can be serious, it has to be assured that the ATCO possesses an unrestricted ability to perform – anything else would be a danger to safety within air traffic control. This requires an integrated approach throughout the whole organization. Managers need to be trained to recognize and handle performance losses, employees need to be informed and convinced that CISM is part of their professionalism, best practice examples should be published in internal papers and professional journals.

Surveys conducted to date have revealed an improved culture of safety for ANSPs with an optimal interaction of ATCOs, peers, and managers. This is also true for an international comparison despite the cultural differences. Moreover, if crises are dealt with professionally, this promotes human factor improvement in general also in other areas of air traffic control. A well-established and solid CISM program positively affects the culture of failure, the culture of safety, the team resource management program, the motivation, and the preservation of work force (Vogt, Leonhardt, Köper and Pennig 2004). A CISM program also has positive economic effects, and the investment in the establishment and the running of the program pays off quickly (see Chapter 12).

Experience

DFS has had wide and positive experience with CISM. It has evolved into a normal procedure, which is offered as a standard by every watch supervisor in charge during a critical incident. This ensures contact between peer and ATCO, which then may be extended to a full CISM intervention, if both parties agree. The initial contact can also just be a bilateral assessment of the situation and the ATCO's responses.

Over 80 certified peers (with all three ICISF courses) are working at DFS. The number of peers at each DFS location matches the size of the unit. Therefore, a peer is either on site or can be called in immediately. The peers are actively supported by the watch supervisors and managers and vice versa. Thus, they can fulfill their task in a professional manner and with the necessary organizational support. Due to the quick availability and the high acceptance of the peers, over 90 percent of DFS CISM interventions are applied as SAFER-R (see Chapter 5), that means immediately and directly after an incident.

As a consequence, more than 90 percent of CI-affected and CISM-treated ATCOs are able to work on the first day of their scheduled shift again. DFS has registered no delayed symptoms or consequential losses ever since the introduction of the CISM program.

Besides the experience of everyday critical incidents, the DFS crisis intervention team has also had experience with major accidents (mid-air collision over Überlingen) and natural disasters (Tsunami). The international cooperation with other ANSPs and the intensified networking within aviation is well-established and ensures mutual support in case of an accident. Since 2000, an annual conference has been conducted at the DFS headquarters to establish and maintain the international ATC CISM network. Current partners in this network are: skyguide, Switzerland; Irish Aviation Authority; NAV Portugal; NAVIAIR, Denmark; ENAV, Italy; LGS,

Latvia; ANS, Czech Republic; Croatia Air Traffic Control; FAA, USA; International Federation of Air Traffic Controllers Associations (IFATCA).

CISM is a standard of care in DFS.

Becoming a peer, having a good education and training and developing a peer personality are a benefit for the organization, the employees and also for the peers themselves. There are a lot of changes in the behavior, in the understanding of others and in the ability for a trained and experienced peer to have social and emotional intelligence. Becoming a peer and being a peer is an important and respectable role to play. It is much more than just applying a method, it is about having a personal outlook and taking responsibility.

Being a peer is a lifetime commitment – as my friend and colleague Preben Lauridsen from the Danish Air Navigation Service Provider NAVIAIR used to say:

> It is like Hotel California, the song by the Eagles, you can check out any time you want, but you can never leave.

References

Leonhardt, J. (2005), Implementation of CISM at the German Air Navigation Service Provider. In B. Kirwan, M. Rodgers and D. Schäfer, *Human Factors Impacts in Air Traffic Management* (pp. 208-226). Aldershot: Ashgate.

Vogt, J. and Kastner, M. (2001), Psychophysiological stress of air traffic controllers. In J. Fahrenberg and M. Myrtek (Eds.), *Progress in ambulatory assessment*, 455 – 476. (Seattle: Hogrefe & Huber).

Vogt, J., Leonhardt, J., Köper, B. and Pennig, S. (2004), Economic evaluation of the Critical Incident Stress Management Program. *The International Journal of Emergency Mental Health* 6:4, 185-196.

Woldring, V.S.M. and Amat, A.-L. (1997), Human Factors Module Critical Incident Stress Management. *HUM.ET1.ST13.3000-REP-01.* (Brussels: EUROCONTROL).

Chapter 8

Critical Incident Stress Management in Airlines

Johanna O'Flaherty

Components of an Airline Crisis Intervention Program

Essentially, there are four components to the airline response-training program:

a) preincident training/mitigation,
b) preparedness,
c) response, and
d) recovery/reintegration (O'Flaherty 2002).

The training consists of organizational structure, which is similar to the therapeutic frame, psychological implications of responding to a disaster, aviation pathology, and the importance of instituting an intentional self-care program including participating in a formal critical incident stress debriefing (CISD). The specific roles of governmental agencies, including the role of the American Red Cross are reviewed. Furthermore, the training covers cultural sensitivities in dealing with death, the needs of caregivers when reintegrating into the workplace. Specific emphasize is paid to the complexities of crisis psychology and particular research is applied when selecting mental health professionals to join the team. Along with the logistical and psychological aspects of crisis response, the role of spirituality in the aftermath of a disaster is explored.

In addition to initial team training and program development, recurrent training on an annual basis, and ongoing training and psychological support for team members and senior management is required. First responders also receive appropriate psychological support at the site. Crisis management programs have a defined structure and when the structure remains strong, it proves to have a positive impact on:

a) performance of First Responders,
b) severity and frequency of psychological, physical, and spiritual symptoms in the short and long-term aftermath of a disaster, and
c) reintegration into the workforce. The research conducted at TWA (1996) post the downing of flight 800 substantiated this hypothesis (O'Flaherty 2002).

Preincident Training Includes an Emphasis on Mitigating the Traumatic Effects of Responding to an Aviation Disaster

Mitigation is the means by which people strive in advance to reduce the effects of disasters. The goal of mitigation is to find ways to minimize injuries from a psychological and physical perspective. Airline employees working as first responders are exposed to acute stress in the aftermath of an aviation disaster. The author's hypothesis is that to avoid (or mitigate) the potential for vicarious traumatization, the following are recommended: careful selection of volunteers, appropriate preincident training, psychological support during mobilization, a proper structure (frame) in place to allow the employee/caregivers to perform their respective duties; and a critical incident stress debriefing prior to reintegration into the workforce (O'Flaherty 2002).

Aviation crisis response programs are obliged to meet the Federal Requirement outlined under Sections 1136 and 41113 of the Family Assistance Act. An abbreviated overview of these programs and the responsibilities of the airline caregivers appear in this chapter. In the aftermath of a disaster, the key to appropriate crisis response is flexibility, but flexibility within a structure. Airline crisis response training focuses on the importance of a systematic approach when responding to traumatic situations.

Preparedness

Preparing for a disaster is the process of planning and preincident training. Activities include emergency planning, ensuring adequate response, and providing adequate support during activation of team members. These may take the form of educational initiatives, simulation exercises, research, and the creation of communication networks for the rapid transmission of information during an airline disaster (O'Flaherty 2002).

Essential Components of an Airline Crisis Response Program

Outlined here are several essential logistical components of a crisis management program, although it is imperative to approach the *psychological* aspect of crisis management as a *work-in-progress*. Conversely, psychological assistance along with debriefing and defusing of the crisis response team is monitored throughout the recovery phase and adapted as needed. Since, first responders are airline employees who have the knowledge and expertise to anticipate probable behavior and emotional reactions on the part of family members; they receive minimal training in the psychological implications of vicarious trauma as the result of responding to an airline emergency. Training focuses on the complications unleashed by a crash, which demands that, the airline employee shift roles from *worker* to *emotional container* (O'Flaherty 2002).

The airline employee sets up the first relationship with grieving relatives and loved ones. Currently, there is limited empirical research on how to perform this challenging task. The families, transported to the site of the crash, are coping with a variety of challenges; they are in a strange environment; strangers and emergency agencies surround them; and they have lost a loved one. Crisis team members must immediately secure a calm environment for the grieving family members. Typically, there are telephones, beepers, two-way radios, and a variety of other noisy technologies in operation, which create a circus-like atmosphere that is not conducive to grieving. At the top of a caregiver's checklist should be the requirement to set up a quiet place for the families to grieve and mourn without disturbance from the chaos of the disaster environment (O'Flaherty 2002).

In addition, airline personnel are usually the first to respond to the site of a major aviation disaster; therefore, this research specifically focuses on examining the effects of such work and the implications for vicarious traumatization among this particular population. The research focuses on preparation, training, and psychological preparedness, as well as the appropriate level of care and support a first responder needs to continue as effective caregivers and employees. O'Flaherty's (2002) research indicates that:

> there is a significant dropout rate for first responders who do not receive appropriate psychological counseling in the *immediate aftermath* of the incident, due to vicarious traumatization (p. 332).

Preincident Training and Development of a Crisis Response Program

If an effective crisis response program is to be successful, the highest executive levels must endorse it. There are several essential points to consider when selecting personnel as members of an airline crisis team. A panel of experienced crisis managers and mental health professionals should interview each potential candidate. The basic criteria for first responders are to possess humility, modesty, and integrity; the candidate must be able to make a good first impression; and be comfortable working with multicultural individuals. Most team members are fluent in at least one other language. This is crucial, as they work with multinationals in the aftermath of an airline disaster. The person must be empathetic while maintaining emotional stability, and able to keep personal projections at a minimum. Above all the candidate must be a team player (O'Flaherty 2002).

Moreover, all candidates must be two years post personal trauma before consideration. At some stage, candidates will be working with bereaved family members, and it is important to minimize counter-transference, which can interfere with the team member's success. The individual's age and life experience also influence the decision-making. If the candidate has had experience in similar situations, naturally that experience is an asset. How other stressful situations were processed can help determine the candidate's general threshold of trauma (O'Flaherty 2002).

Once selected, a team member will participate in an initial three-day training seminar, which covers the logistics of crisis management, psychological aspects, cultural sensitivities, and self-care. Due to the graphic portions of the training, candidates' unresolved emotional issues could surface; therefore, they may withdraw from training until a later date (O'Flaherty 2002).

Assisting Victim Survivors and Families is the Role of the American Red Cross

While assisting (victims, survivors, and families) is not the primary function of the internal Employee Assistance Professional/Mental Health Professional (EAP/MHP) counselor, it may be necessary to provide interim support services to these individuals until emergency mental health professionals arrive at the scene. Clinicians will have to use their own judgment and *do the right thing*.

Response: Emergency Levels and Phases of Response to an Aviation Disaster

A *critical incident* is an event that is so unusual and extraordinary that it threatens to overwhelm people's natural ability to cope with difficult situations. The crisis manager, upon notification of the incident, determines the level of response required. Within the airline industry, there are six standard emergency levels, and team members respond to most of these emergencies (see Table 8.1).

Table 8.1   Emergency levels within the airline industry

Level 1:	Any critical incident that has affected an employee or employee group
Level 2:	Loss of life of a passenger
Level 3:	Sudden accidental loss of life of an employee
Level 4:	Loss of life of an aviation employee, death in the line of duty
Level 5:	Serious incident but no fatalities
Level 6:	A catastrophic emergency where multiple deaths have occurred

Recovery: The Airline Crisis Team's Role

Recovery occurs after addressing peoples' immediate needs. Within the first 48 hours, all team leaders must have contacted the activation line and be prepared to mobilize at the direction of the crisis manager. Some team members will already be in transit to the Accident Administrative Center. The Command Center has been established and has prepared for data collection and input, including such information as, which team members have been activated?

Rendering support, accurate information, and guidance are by far the most important tasks in the assistance after the crash. Information gathering becomes the primary focus of family members in their efforts to cope with the news. In the first hours, the only individuals that have access to the data are the aviation staff. Selectively and carefully sharing information with family members is essential in helping them learn to cope (O'Flaherty 2004).

The news networks are competing for *breaking news*, and reporting can be very aggressive. The journalist's job is to report news, but in the immediate aftermath of a disaster, when information is sketchy, reporters tend to follow grieving family members and literally shove cameras in their faces. First responders can be invaluable in protecting grieving family members from the media.

Once assigned to families, each team member meets their respective family at a designated location and escorts them to the site (hotel) where family members are congregating. The team members ensure the provision of all lodging and other immediate needs of the family. Essentially, the team members stay with the family for the duration of the immediate aftermath. They accompany the family to the daily briefings conducted by representatives from the airline, the NTSB, and the Medical Examiner's office. These briefings are often very graphic and, therefore, painful for the family – and the team member (O'Flaherty 2002).

The following outline illustrates the five phases of response and organizational structures for crisis management. When an accident occurs that involves casualties, the structure includes the following.

Phase 1: Mobilization of the Team

Activate team members and advise them of their assignment. Set up a joint *Family Center* with an operational working space for the airline first responders along with the Red Cross, National Transportation Safety Board (NTSB), and other governmental agencies. Identify a clear command structure at the scene while working with intra-agency and interagencies.

Ensure that responders are cognizant of role definitions and boundaries. Have a mechanism in place to *resolve conflict*, because in a crisis everyone regresses psychologically – there will be conflict! Maintain the supportive structure, such as safe logistical and psychological *containers* so the team can carry out their duties. Provide on-going support, including access to daily defusing, psychological assistance, and an exit debriefing for all first responders. Set up a quiet place for spiritual reflection, clergy from the respective religions of the bereaved along with first responders usually staff this room.

Phase 2: Stabilization

At this stage, triage is resolved. The injured are in the hospital (intensive care unit or burn center) and *stabilized*. Team members/first responders begin the process of securing dental records and medical x-rays for the coroner. If the accident occurs overseas, caregivers may assist with the completion of victim identification forms. Crisis team members are aware of cultural sensitivities and assume a subservient role to local law enforcement and other agencies in the country where the incident occurred. The local coroner has begun the painful process of identification and repatriation. First funeral and/or memorial service takes place. Throughout the crisis response, ongoing psychological support is provided for all caregivers.

Phase 3: Demobilization

Demobilization is a particularly difficult phase for the caregivers; the NTSB and other respective governmental agencies are the ones who determine that it is time to close the family center. The remaining family members, along with the caregivers, return home. Sometimes family members must return home without their loved one's remains; this is heartbreaking for all involved. All caregivers are scheduled for a Critical Incident Stress Debriefing (CISD). Therapeutic time off is necessary, as team members need to integrate the experience before returning to their regular airline jobs. Duration of time off is based upon the individual's duties and the time involved as a caregiver.

Phase 4: Corporate Recovery

An administrative debriefing occurs as soon as possible. It concentrates on *lessons learned*; blaming and *second guessing* are counter-productive. Part of the corporate recovery is for executives to participate in a clinical debriefing. It is advisable to have a contract clinician who is familiar with crisis work, CISD, and the airline industry to facilitate the executive debriefing, to minimize projections, enmeshment, and further complications for the internal EAP. The incident is acknowledged in internal communications. All employees are recognized and thanked. *Reminder*: In the aftermath of a disaster, a *business as usual* approach is insensitive and counter-productive. Employees are integral in helping the company to recover from a crisis. Therefore, a careful sense of timing and sensitivity is called for in the aftermath. Attention should be focused on the wounded psyches' of airline employees, so that they do not feel discounted as the company seeks recovery. Ongoing psychological support for corporate executives is a requisite component of crisis management. The internal Employee Assistance Professional (EAP) can facilitate this function. If the company does not have an EAP, the assistance of an external provider should be enlisted.

Phase 5: Community Recovery

Employees have returned to their regular duties. Memorial services for the unidentified take place. This is extremely disturbing, as the remains have not been recovered. Essentially, each family is assigned a funeral director naturally the family's culture and customs are honored. A symbolic burial takes place, which means that an empty casket is buried. Psychological complications occur for the family and caregiver when remains are forensically identified post the symbolic burial. Usually, the grave is reopened and the remains are buried or cremated and the ashes are spread over the grave. Preparations for the first anniversary of the tragedy begin. This is a joint effort with representatives from the Family Advocate Group, the Airline, the Red Cross, and the NTSB. The Family Advocate Group oversees the construction of appropriate monuments. Throughout the recovery phases, it is imperative that careful attention is paid to cultural needs.

The Role of Employee Assistance Programs and Mental Health Personnel in the Aftermath of a Disaster

The Employee Assistance Counselor or contracted Mental Health Professionals (EAP/MHP) have an excellent vantage point for observing the intricacies of the operation. If these professionals notice anything out of the ordinary concerning an employee's behavior, or anything that seems concerning, they should bring it to the attention of the person directing the mental health efforts. Examples include caregivers who are inappropriately dressed or those who avoid rest breaks. Additionally, if the caregiver exhibits a diminished or absent need for food, resists demobilization, or has a tough time detaching from their duties. The EAP/MHP counselors are advised to pay attention to the team members that are isolating and not engaging with their support organizations. If the EAP/MHP needs to remove a team member from the scene, it is imperative to send a companion with the team member. Additionally, the EAP/MHP may need to reassign team members to lighter or alternative duties. Another red flag issue is team member's over-involvement with the family that they are assisting.

Employee Assistant Professionals: Advice and Counsel to Senior Management

Here are some suggestions for advising management on the appropriate care of employees involved in critical incidents. EAP counselors may offer advice and counsel to management when appropriate, but they have no management or command authority. All decisions are the responsibility of the appropriate management representative. EAP counselors do not assume a position of leadership or authority for incident management. They *advise* management on minimizing stress effects, maximizing performance, and facilitating overall corporate recovery from a psychological perspective.

As soon as the level of emergency has been determined, the EAP coordinator assigns schedules for the EAP counselors to be available for defusing and debriefings. Office staff are responsible for handling the telephones and setting up offices to accommodate the team members' activities. A note of caution for the debriefing counselors: do not interrupt the response teams' process. Allow the teams to complete their respective duties before scheduling debriefings. The office staff keeps a communications log of incoming questions and answers, and sets up a message center for counselors and team members (O'Flaherty 2002).

One-on-One Interventions

Workers who display obvious signs of distress may avail themselves of one-on-one interventions (see Chapter 5). The therapist need to pay attention to signs of obvious distress such as crying, a shock-like state, a fear of flying, unusual behavior (such as, change in cognitive skills), *acting out* (for example, punching and screaming). Naturally, all interventions occur when personnel are not actively engaged in service.

Ideally, all persons receiving one-on-one intervention are given an additional rest period after the intervention is completed. During the rest period, the counselor remains in touch with that person. The counselor tells the team member that he or she will be back to check on their progress within minutes, and requests that the team member remain in a resting position until then. Upon returning, the counselor should determine, whether the worker can return to duty or be given an alternate assignment. If the worker continues to be distraught it is wise to have medical personnel assess the vitals of the distressed individual and evaluate them for injuries.

Defusings and Debriefings

A defusing (see Chapter 5) is a small group meeting conducted within hours of the incident. Defusing takes place after the event with the personnel directly involved in the incident. The purpose is to offer information and support, allow initial venting of feelings, set up or establish whether there is a need for formal debriefing (see Chapter 5), and to stabilize team members so they can go home or back into service. A defusing is similar to a *mini debriefing*, but is not as detailed or as long.

The following criteria for a formal Critical Incident Stress Debriefing (CISD) was formulated between American Airlines Employee Assistance Program Counselors and the flight attendants union Association Professional Flight Attendants (APFA).

Table 8.2 **Critical incident stress debriefing criteria formulated between American Airlines Employee Assistance Program counselors and the flight attendants union Association Professional Flight Attendants**

Level 1:	Aircraft accident/air disaster
Level 2:	Death in flight or on a layover
Level 3:	Evacuation using slides
Level 4:	Fire on board
Level 5:	Flight attendants physically assaulted during a flight or a layover
Level 6:	Hijacking
Level 7:	Medical emergency in flight
Level 8:	Severe turbulence with or without injuries
Level 9:	Terrorism

Guidelines for Defusing Services

1. Defusing should be done within hours of the critical incident. The ideal period is 3-4 hours post-incident, at the end of the same day. If it is not possible to hold the defusing within this period, a formal debriefing will be necessary. The key is immediate intervention.
2. Defusing is a group process (as opposed to one-on-one), and all persons involved in the incident should attend the defusing.
3. Defusing lasts approximately 45 minutes.
4. Peer Support Persons can perform defusing, but should be well aware of their personal limitations and should call for support from a Mental Health Professional if the situation warrants. Peers directly involved with the operations should not perform defusing for Crisis Response Team members.
5. Defusing should be held in a comfortable atmosphere. All parties should remain in attendance until the conclusion.

There are three essential components to a defusing 1) introduction, 2) exploration, and 3) information. The format for the defusing shall be as follows:

1. Introduction: Facilitator states purpose of meeting and sets guidelines establishes a safe container. Confidentiality is emphasized; the team leader describes the process, and offers additional support wherever needed.
2. Exploration: Team leader asks the group to tell you what happened and asks a few clarifying questions such as asking the group; what was the worst/most disturbing part? Encourage participants to share experiences and reactions. Allow freedom of discussion to take place on the *worst/most disturbing part*. After the discussion subsides, offer information on possible signs and symptoms of stress they may or may not have experienced also provide information on what they can do about it. Initially, allow ventilation of

feelings. Acknowledge the feelings, validate the feelings, and move on. DO NOT probe or dwell, as it is much too early after the critical incident for this tactic. Team leader is constantly assessing the participants to observe who needs assistance that is more psychological.
3. Information: Team summarizes the participant's exploration. Facilitators begin the process of normalizing group's experiences and reactions. The information phase is a teaching phase therefore, facilitators provide an informational handout to each group member. It is essential that group members know how to contact each other. Additionally, it is crucial that the participants be given a list of mental health professionals who are specifically trained in crisis psychology.

It is advisable to keep the defusing-session informal, but to the point. Do not allow the defusing to lapse into a *Critique of Operations*. The CISM team's primary function is to facilitate and direct the session. If the team leader is a mental health professional, this is an opportunity to assess the need for a formal debriefing and schedule an appropriate time for the CISD.

Critical Incident Stress Debriefings Defined

From a depth psychological perspective a CISD (see Chapter 10) provides the opportunity for the therapist to create a safe therapeutic container (albeit, it is recognized that CISD is not group therapy). The CISD and defusing processes may be defined as group meetings or discussions about a traumatic event, or series of traumatic events. The CISD and defusing processes are solidly based in crisis intervention theory and educational intervention theory. The CISD and defusing processes are designed to mitigate the psychological impact of a traumatic event, prevent the subsequent development of post-traumatic syndromes, and serve as an early identification mechanism for individuals who will require professional mental health follow-up subsequent to a traumatic event. Everly and Mitchell (1983) developed this seven step model of debriefing primarily for first responders such as para-medics, fire fighters, police, and other law enforcement agencies. This model of early intervention was modified and implemented in the airline industry in 1990; CISD has proven to be an effective modality of early interventions.

There are seven specific stages to a CISD, which are facilitated by a trained peer and a Mental Health Professional (or two Mental Health Professionals). This procedure usually takes about three hours. If the debriefing is facilitated for law-enforcement due to a line-of-duty-death, it is suggested that the chaplain be present at the CISD. The debriefing is re-structured and the teaching phase is used to plan for the funeral. The seven core phases of CISD are as follows:

1. Introduction: To introduce the facilitators and participants. The team leader, who is a Mental Health Professional, will explain the process, and set

expectations. Confidentiality is emphasized and the participants are assured that there will be no notes taken during the debriefing. The participants are asked to provide their contact information for follow- up, a copy is given to each participant at the conclusion of the debriefing. The participants are encouraged to stay in contact with each other.

2. Fact Phase: The traumatic event is described from each participant's perspective on a cognitive level. Essentially, this is the time when the traumatic event is weaved and the trauma story is told. This phase is crucial as the participants will hear the event being described from different perspectives. Airline crewmembers who experience an unusual event on-board a Boeing 777 (wide-body aircraft) will have varied versions of a similar event. For example, the flight attendant working in the first class cabin will have had a different experience to the flight attendant working the aft galley. And, the pilots' experience will be completely different. Due to the individuality of each participants' experience, it is important that all crewmembers are encouraged to attend the debriefing. The majority of major airlines adapted the Everly and Mitchell (1983) model of CISDs as a component of their overall crisis management programs in 1990.

3. Thought Phase: This phase provides an opportunity for the participants to describe the event on a cognitive level and to transition to emotional reactions. The team leader will invite the group members to share their most prominent thoughts about the event.

4. Reaction Phase: Participants are asked to identify the most traumatic aspect of the event, from their perspective, and to try and connect to an emotional reaction. This is an opportune time for the Mental Health Professional to facilitate an emotional catharsis by continuing to provide the appropriate frame/container.

5. Symptom Phase: Participants are asked to identify personal symptoms of distress from a cognitive, emotional, physical and spiritual aspect. At this juncture, the facilitators will begin the process of transitioning from the emotional level to the cognitive level.

6. Teaching Phase: This phase provides an opportunity for the facilitators to educate on the normal reactions to traumatic experiences. The participants are encouraged to practice healthy stress management techniques, such as exercise, a balanced nutritional diet, and it is crucial that the participant is able to identify a support system prior to concluding the debriefing. The teaching phase of the debriefing concentrates on adaptive coping mechanisms, by providing comprehensive suggestions on self-care practices from a psychological, physical, emotional, and spiritual perspective. Special attention is given to outlining the tendencies towards maladaptive coping mechanism such as isolating, self-soothing by drinking excessively or self-medicating by the inappropriate use of over the counter medications (usually sleeping aids) or, indeed, prescription drugs. The participants are advised to seek professional consultation from their physicians if they experience any of

the signs and symptoms of Post Traumatic Stress Disorder discussed later in this chapter.
7. Re-Entry Phase: At this phase, the facilitators will synthesize any themes that occurred during the preceding phases and encourage participants to clarify any ambiguities; as they prepare for termination of CISD. Follow-up sessions, if needed, will be scheduled at this time. At the conclusion of the CISD, refreshments are served – this is similar to the Native Americans symbolic ritual of *breaking-bread* after a sweat lodge. Metaphorically, the group experienced an emotional/spiritual sweat lodge as they have shared on a deep emotional and spiritual level.

Although, the debriefing process has both psychological and educational elements, it should *not* be considered psychotherapy. Instead, it is a structured group meeting or discussion in which personnel are given the opportunity to discuss their thoughts and emotions about a distressing event in a controlled, planned, and rational manner. Analogous to a safe therapeutic container. They also get the opportunity to see that they are not alone in their reactions but that many others are experiencing the same reactions (For a complete guideline on debriefings, see Mitchell and Everly 2001).

Post Traumatic Stress Disorder

An important element of all CISD debriefings is to educate on signs and symptoms of stress and PTSD (see Chapter 3 and Chapter 4). With this disorder, the person has experienced an event, which, is outside the range of usual human experience that would be markedly distressing to almost anyone. Examples of such experiences include serious threat to one's life or physical integrity, serious threat or harm to one's children, spouse, or other close relatives and friends; sudden destruction of one's home or community; witnessing a serious injury or fatal accident or physical violence.

Van der Kolk et al. (1996, p. 91) state that traumatic events are re-experienced persistently in at least one of these ways: Recurrent, intrusive, and distressing recollections of the event (in young children, repetitive plays in which themes or aspects of the trauma are expressed). Recurrent distressing dreams of the event, sudden acting or feeling as if the traumatic event were recurring (includes *reliving* the experience, illusions, hallucinations, and dissociative episodes, even those that occur upon awakening or when intoxicated. Another example is one of experiencing intense psychological distress at exposure to events that symbolize or resemble an aspect of the traumatic event, including anniversaries of the trauma.

Other indications of trauma are when the individual persistently avoids stimuli associated with the trauma, or exhibits numbing of general responsiveness (not present before the trauma), as indicated by at least three of the following: Efforts to avoid thoughts or feelings associated with the trauma, efforts to avoid activities or situations that arouse recollections of the trauma, inability to recall an important aspect of the trauma (psychogenic amnesia), and markedly diminished interest in significant activities (children, may experience a loss of recently acquired developmental skills, such as, toilet training or language). Other symptoms include feelings of detachment or estrangement from others, restricted range of affect, for example, unable to feel love, a sense of foreshortened future, including no expectation of having a career, marriage, children, or a long life (van der Kolk et al. 1996, p. 92).

Further symptoms include: Newly presented symptoms of persistent increased arousal (not evident before the trauma), as indicated by at least two of the following: Difficulty falling or staying asleep, irritability or outbursts of anger, difficulty concentrating, hyper-vigilance, exaggerated startle response. Physiologic reactivity upon exposure to events that symbolize or resemble an aspect of the traumatic event (for example, a woman who was raped in an elevator breaks out in a sweat when entering an elevator). If any of these symptoms persist beyond three weeks, the individual is advised to contact a mental health professional and/or seek medical attention from their family physician. At the conclusion of this phase, it is imperative to provide a cognitive anchor for the participants.

Confidentiality and Debriefings

In addition to good practice, confidentiality is also governed by several legal principles. It is unlikely that anyone on a Critical Incident Stress team would deliberately spread untruthful statements about someone in a debriefing session. However, if an innocent remark concerning someone in a debriefing gets turned around in the translation, you may find yourself involved in a legal action. The best defense is to *simply to keep everything confidential.*

Subsequent Investigation Testimony

In the aftermath of an aviation disaster, closely tied to the topic of confidentiality is the topic of subsequent investigation testimony. It is quite possible that civil and/or criminal litigation will arise out of an incident you are debriefing, investigators such as the NTSB and FAA may want to talk to you or obtain copies of your records. Plaintiff's lawyers, defense lawyers, district attorneys, and investigators may seek out debriefers.

Additionally, a judge may command that the facilitators release records of the debriefing, therefore, there are no notes or records taken during a debriefing. Consequently, the facilitators will advise the participants not to disclose any technical information that may incriminate them; they are reminded that the CISD is not a critique of standard operating procedures. In addition, the fact phase (where possible incriminating information may be disclosed) is only a small portion of the process. The facilitators are mainly concerned with the participant's emotional reaction to an event.

The facilitators need to advise those being debriefed ahead of time, that they need not say anything that they feel could compromise an investigation or incriminate themselves. This is *especially important for airline pilots*; the peer-pilot (in the debriefing) will advise their colleagues of any potential incriminating disclosures (ALPA 2001). The participants need to be advised that confidentiality in a group debriefing setting cannot be guaranteed, if facilitators are mandated by a court order to testify. The facilitator's task is to skillfully direct the interventions towards the emotional and reaction phase of the debriefing. It is wise to remember that a debriefing *is not a critique of the operations*. Crewmembers are assured that a debriefing is not a company, Federal Aviation Administration or National Transportation Safety Board critique, rather the debriefing is facilitated to normalize the experience and concentrate on the emotional well-being of the crewmembers.

It is advisable for the mental health professional as the team leader to investigate the existence of client/patient privilege laws in the particular state where the debriefing is being held. Most states provide for the confidentiality of statements made to a mental health professional while undergoing treatment. If these laws don't apply to group settings, such as debriefings, the therapist may want to direct someone who wants to *disclose*, to a private session with a mental health professional who is licensed in that state.

First Responders Reintegrating into the Workplace

Employees who are directly involved with the traumatic event, especially if death has occurred, must be given appropriate time off to attend memorial services. In addition, it is essential that coworkers be briefed on the sensitivity of work-related trauma and the challenges of reintegration into the workforce. A collaborative effort between senior management and employees, regarding decision-making in respect to responding to trauma, is warranted. Decisions made without the input of those involved can have negative consequences, such as, "You weren't there, yet you're making decisions without my input." When personnel receive the appropriate support, they feel validated and acknowledged as an integral part of the *airline family*. Expressions of appreciation are very important and meaningful. In the aftermath of Flight 800, we gave small crystal angels to employees. This minor gesture of acknowledgement was greatly appreciated; to this day, the employees treasure these angels. During a crisis in the workplace, employees need to feel that they are more than a number and that their contribution is valued beyond the production line.

References

Airline Pilots Association ALPA (2001), Unpublished research. (Washington D.C.).
Everly, G.S. and Mitchell, J. (1983), *Innovations in disaster and Trauma psychology*. (MD. Ellicott City: Chevron).
Mitchell, J.T. (1995), Traumatic stress in disaster and emergency personnel: Prevention

and intervention. In J. P. Wilson and Beverly Raphael (Eds.), *International handbook on traumatic stress syndromes*. (New York: Plenum Press).

Mitchell, J.T. and Everly, G.G. (2001), *Critical Incident Stress Debriefing. An Operation Manual for the Prevention of Traumatic Stress among Emergency Services and Sisater Workers*. (Ellicott City, MD, USA: Chevron Publishing).

O'Flaherty, J. (2002), *The aviation industry, occupational medicine*. (Philadelphia, PA: Hanley & Belfus, Inc.).

van der Kolk, B.A., McFarlane, A.C. and Weisaeth, L. (1996), *Traumatic stress: The effects of overwhelming experience on mind, body and society*. (New York: Guilford Press).

Chapter 9

Critical Incident Stress Management at Frankfurt Airport (CISM Team FRAPORT)

Walter Gaber and Annette Drozd

FRAPORT AG, the owner and manager of Frankfurt Airport, has expanded its emergency management system with a component for crisis intervention and so called *psychological first aid*.

The counseling service that is being formed will support the responsible persons in emergency management and disaster relief management to cope with crisis and emergency situations by providing qualified and competent psychosocial *first aid* to affected persons, relatives, and helpers.

In recent years many German Rescue Services have developed activities to ensure professional treatment of traumatized persons during a current problematical situation in life.

During all major national and international catastrophes in recent years the press media has always reported on the counseling provided by the *professional helpers* (fire department, police, medical personnel) and have also demanded professional counseling and guidance (psychological first aid) for helpers spending so much time at the site of the catastrophe.

This was organized in the end (examples: aircraft accident in Ramstein in 1988, ICE train accident in Eschede in 1998, Concorde accident in Paris in 2000).

FRAPORT AG, as operator of the airport, is responsible for the safe operation of the airport without disturbances.

Consequently, an organized emergency management system must be set up – in addition to the present service provided by the experts of the medical services – which can go to work immediately in the case of a so called major accident or catastrophe.

This applies for the case of such an accident/catastrophe occurring at the airport or in the immediate vicinity (so called airport-related events) or further away (so called flight-related events).

Special Requirements for Airports

If one intends to successfully establish a CISM Team at an international airport it will be important to consider a number of different interfaces that will decisively contribute to the success of such a team.

- Approx. 52,000,000 passengers in 2005
- Direct link to two train stations (local trains and high-speed long distance trains)
- Two major highways in direct vicinity
- 14,000 employees at FRAPORT AG
- Over 60,000 employees at Frankfurt Airport
- 500 companies operating at the airport
- Employees from 30 different nations work at the airport

Figure 9.1   Frankfurt Airport traffic data

Figure 9.1 gives some key numbers of Frankfurt Airport as an example, which had to be considered in the planning and implementation of a CISM program.

Responsibilities

At an early stage it must generally be clearly defined who is in charge of what task and the resulting responsibilities. The airport (as a service-provider) – if not otherwise agreed by contract – is not responsible for passengers, meeters, customers and employees of other companies. In other words, the responsibility is in the hands of external persons.

Meeters/greeters and media representatives are generally requested to turn to the airport operator.

An airport operator is well advised to set up internal structures to appropriately handle incidents with varying dimension. Based on image questions and media reporting in recent years, the airport operator will be heavily focused on by media reporters.

Cooperation agreements with airlines and their care teams serve to have a defined interface which will help ensure good cooperation in handling a possible incident.

Airlines

National and international airlines must stick to the instructions of the home base and will possibly only cooperate to a limited extent. In addition, they will exclusively keep the interests of the airline in mind.

The cultural particularities of various people from different countries at an international airport must always be inquired about and considered.

Current political events (for example wars) must always be taken into consideration when helping affected persons with different cultural backgrounds.

Depending on the *size* of the airline, there will be several representatives and in some cases no (!) representative will be able or be authorized to make decisions.

This means that in a case of doubt the representatives will make decisions after having consulted the home office which will not always benefit the affected persons (for example no, or lack of, communication with family members).

Considering the above, there is an absolute necessity to discuss this topic before the occurrence of a major incident and to define interfaces in agreements to ensure that responsibilities are assigned beforehand.

At FRAPORT AG, the Airport Operational Committee (AOC) regularly holds meetings and exercises with the responsible persons. This ensures that *key players* get to know one another and a trustful working relationship is established.

Meeters/Greeters

Should there be a catastrophe, we must generally expect a large number of meeters and greeters because the airport will always be the place to go to if there is an airline accident.

This means that at least three times the number of people (900 to 1,200 persons per aircraft) will have intensive emotions in different constellations at any time of the day or night at the airport.

These persons require the care of a continuously operating care team in order to keep these persons informed as best as possible.

They must be isolated from the press and other persons at the airport in order to be provided with *bad news* or to be joined by their family members.

Helpers having to give *bad news* must be trained for this. Furthermore, a sufficient number of helpers must be on hand.

Facilities/rooms and catering – after having clarified the responsibility for this – must be made available to meeters and greeters free of charge.

Different Religions

In this day and age of globalization we are confronted with a multitude of religions which must be appropriately taken into consideration.

This means early cooperation with the representatives of various religious congregations in the area/region and to better acquaint them with the specifics of an airport. There have been contacts with representatives from various religious groups for years now and their names are on file in the Crisis Management Center and at Medical Services.

Some religious representatives have already set up a hotline and a 24-hour *on call* service. This has proved to be useful in the past. For example, during Christmas Eve 2005 when a FRAPORT employee committed suicide.

Care and the notification of death was given by the Protestant pastor and a representative of the company. CISM care for the colleagues was requested and ensured by the doctor in charge at the airport on 24 December 2005 at about 2200 hours.

Regular meetings at the airport (invitation comes from FRAPORT AG) have proven to be a success for many years now. "One gets to know one another" is the most frequent statement and this is important in order to assist one another.

Different Nationalities

In the case of a catastrophe at the airport or due to an airline-related accident at the airport or in the vicinity of the airport, we must generally assume that many persons with different nationalities will be affected.

The different cultural backgrounds will play a major factor in dealing with the problems (screaming, grief, persons collapsing, blaming others, aggression against persons or items) and there is no telling what will happen in advance. It is a big asset of an international airport to be able to utilize its workforce which is also made up of various cultural backgrounds.

Even if these colleagues are not already members of the care teams, their presence alone would be of major assistance in dealing with affected foreigners because by translating or helping with minor tasks the grief and suffering of affected persons could be minimized to a greater extent. Lists of the names of foreign employees with their cell phone numbers should be on file.

24 Hour Shift/Shifts of Helpers

When setting up a team of helpers at the airport, one must make a clear decision in regard to how many employees per shift should be at the airport and thereby take the shift operations at the airport into consideration.

As early as possible, one must attempt to estimate the time frame of the work needed of the helpers in order not to request all (!!) helpers to come to the airport at once. A reserve team must be ready to go to work in order to replace some colleagues or take the place of an already working team that is exhausted.

As a general rule about 30 per cent of the helpers should be kept in reserve and not be involved in the *hot phase* of the work to be done.

Contacting/Commuters

Many of the helpers are commuters.

This means that they are dependent on car pools or it is not possible to precisely determine how long they will be able to help during the early phase.

The responsible team managers must also determine at an early phase how team members can get back home safely after they have worked.

It may be necessary to offer team members a taxi home free of charge or have them driven home by company drivers. Such matters must be clarified as soon as possible to ensure that the helpers can best concentrate on their work and duties.

Catering/Relaxation Room

Catering is an important element for the affected persons and the helpers, too. Free of charge catering must be ensured for all members of the help team in a separate room so that the individual can relax for a limited period of time, talk with other helpers and have a *cushion* to lean back on.

A room for relaxing in total silence should also be provided if the work of the help team is expected to last longer than six to eight hours.

Multitude of Companies

The multitude of companies, which operate at the airport or in the vicinity, offers great possibilities of cooperation in many sectors such as logistics, communications, catering, rooms/shelter, transportation capacities. Here cooperation will be needed and should be initiated.

Government Agencies (Customs, Police, Federal Police)

After a catastrophe several government agencies will rush to the scene with different responsibilities. It is not always clear who has the *primary* authority to do what, specifically during the first hours after the catastrophe.

For example, approximately 14,000 passengers returned from Asia at the end of 2004/early 2005 (tsunami victims) within seven days. Many of them had no money, no clothing, no documents and were in a traumatized state.

They understandably had little understanding for questions from government authorities such as:

- Do you have something to declare?
- May I see your passport?
- Where are your family members?
- Who will take care of you?
- Where do you live in Germany?

Here, too, improvements are being made by regular meetings and exercises involving key players of the airport operator and key players of the government agencies to establish better cooperation and trust.

It was also noticed that representatives of government authorities and agencies cooperated very efficiently with airport personnel and did everything in their power to help the affected persons as best as they could.

For example, many tsunami victims received new temporary identification cards immediately, the customs formalities were reduced to the absolute minimum, airlines gave out tickets for the victims to continue travel by rail or handed out hotel vouchers. A special care team of one airline started immediately taking care of smaller groups of affected persons.

What Can our Team Expect in the Case of a Catastrophe?
(number of flights after 2004 tsunami)

It is very important to immediately estimate how many affected persons and, subsequently, how many meeters/greeters we can expect.

- What is our current manpower?
- How much can we take on at maximum with the existing team?
- When will we need our backup team?
- Must we activate a crisis team?
- Our experience tells us that at maximum two flights can be handled parallel and then the capacity limit will be reached.

Handling two flights means the immediate handling of up to 600 passengers and approximately 1,800 meeters/greeters (above-mentioned one to three ratio).

Backup Team (Cooperation)

A catastrophe at an airport will always mean the handling (providing assistance) of many affected persons and a high number of meeters/greeters.

Good and efficient cooperation with partners in the airline business (airlines, airports, air traffic controller, airport pastors, etc.) will be a top requirement in order to tackle such a catastrophe and thereby also assist the helpers in the best possible manner.

In the opinion of the author, external helpers without any *airline-related experience* (usually these persons volunteer their help) should not be included in the mission for security and operational reasons.

Confidentiality

Confidentiality is a very important factor for persons working in help teams.

The affected persons must always have the feeling that personal data and statements will never be passed on to friends, interested parties or media representatives.

Consequently, the confidentiality factor should always be raised at the briefing session before the start of a mission.

Should anybody on the team disregard the confidentiality rule, the team manager should remove this person from the team. This should also be done to protect the team.

Compensation/Pay

Last but not least this topic must also be discussed. Training expenses for team members are most usually paid for by the company. During a mission the team members should be paid as they would be paid for a normal day/night at work.[1]

At our company, a member of the board of management will come to a meeting (small party) of all helpers a few weeks after the mission has been completed and express the gratitude of the board of management and thank all persons involved.

Checklists

Upon a catastrophe several airlines and government agencies at the airport will become active immediately. Thus there is an absolute necessity for structured procedures. We have set up a checklist which must be discussed, modified and agreed upon by all.

An airport representative should take charge here but only function as a *coordinator*. Responsibilities are at the involved government agencies and airlines.

Experience gained over the past ten years tells us that up to 20 different *key players* may be involved with varying interests which have to be coordinated. Individual actions or measures – without coordination – must be prevented by all involved persons because they are not productive and often harm the mutual interest of involved parties to handle the catastrophe the best possible way.

Checklist

Subject	Comments	Done (?)
Who is responsible? We? Airline? Gov. Agency?	• Are we responsible? • If not, who is our contact partner? • What is expected from us? • Who bears the expense?	

---

1   Personal note of the authors of this chapter: We are of the opinion that no extra pay or compensation should be awarded for this highly stressful work.

Alarm	• Is my team informed? • Who coordinates? • When, what feedback? • Backup team!! • (do not call in but inform)
Religion	What religions will we probably deal with? Are the various religious representatives already informed in advance?
Catering	Has sufficient catering been ordered for • affected persons • employees • team members food warm/cold beverages warm/cold sweets 24 hours/365 days!!! cell phone numbers of responsible persons
Special parking position for aircraft	Has the aircraft been assigned a special parking position?
Clothing	Sufficient amounts? • Shoes • Jogging suits • Shirts • Clothes for children
Transportation	Transportation • Aircraft – *special rooms* • Meeters/greeters of passengers • Team members

Rooms/Facilities	Separate rooms/facilities for • Meeters/greeters • Affected persons • Handling the groups of persons
Handling (Taking Care)	Meeters/greeters Passengers Team members Others e.g. Airline members • Check-in • Ramp • Terminal • Tower
Government Agencies	Contact partners • Customs • Police • Federal Police • Tower • Security • Department of Health
Other	• Taxi vouchers • Hotel vouchers • Passport / ID card (temporary) • Telephone vouchers • Train tickets • Coordination meeters/greeters
Care Team	Ensure taking care of helpers after end of shift; nobody leaves the scene without having been taken care of
Cooperation	Ensure cooperation for setting up a qualified backup
Quality	Regular training and thus ensure international standard according to Mitchell

Implementation of CISM at FRAPORT

For dealing with traumatized persons we intend to train personnel working in security and emergency management (security operations center, airport security, technical operations center, fire department, medical services, operative units) and personnel from other operative and administrative units of our company.

Trained employees can (partially, temporarily) provide assistance to the professionals in the rescue and medical fields in case of an emergency event.

The trained employees can be made available for external customers (airlines, passengers) as well as for internal employees and colleagues as contact partner or counselor.

> Knowledge of the procedures, facilities, responsible persons, work areas, etc. in the own company – like an international airport – are a big advantage.

Experience in mutual work areas with external customers and specifically airlines (DLH, Delta Airlines, United Airlines, Air France) is helpful as well as the Catholic and Protestant pastors working at Frankfurt Airport.

Focus of training measures is also on sensitivity and developing skills of employees to be moderators for internal debriefings after stressful work or missions connected to *critical incident stress management* according to Mitchell and Everly (CISM).

The Medical Services unit at Frankfurt Airport has already conducted several training sessions based on the Mitchell concept for their employees (doctors, emergency doctors, nurses, medics, psychologists, drug counselors) in close cooperation with DFS/ATC (German Air Traffic Control) in Langen/Germany.

The training sessions take place – with focus on a possibly extensive network generation – together with managing supervisors of the company fire department, emergency management and in cooperation with employees of DFS/ATC.

Special stress for our employees was noticed in the past upon the occurrence of fatal accidents of passengers in the terminals, life-threatening and sudden medical problems, unsuccessful reanimation, suicides, fatal accidents on the apron, fatal vehicle accidents of colleagues.

In dealing with affected and stressed employees and colleagues it is of major assistance if experienced and trained colleagues (peers) can speak openly about what has happened.

This experience is confirmed by other counseling teams (for example by German Rail). The offer of a colleague to help in a situation is often accepted much easier than counseling by a person unknown to the affected person.

The training together (as described above) does not only enable getting to know one another better.

Furthermore, there is the possibility to learn about the individual responsibilities and job functions of the colleagues, the necessity for certain procedures and trustworthy cooperation in the twilight zones of the own ability to help.

The limits of the own abilities can be noticed and accepted much easier in such a group.

The implementation of subsequent mission briefings, group discussions, debriefings – whereby taking the own stress under consideration – is a significant step in a medical sense to ensure primary and secondary prevention of stress disturbances or posttraumatic reactions. These prevention services are available to colleagues in the emergency management center, call center and other information departments.

- Information meetings
- Personal and confidential talks, if requested
- Selection procedure by individual interviews and group talks
- Two days training with internal and external psychologists (with experience in therapy for posttraumatic stress disturbances)
- Refresher courses (1 day)

Figure 9.2   Training modules for Crisis Intervention Team

Development of Counseling Services

Not only since 1996 have there been situations that required the work of the counseling team.

Table 9.1   Examples for missions of the FRAPORT Medical Team providing counseling for victims of accidents and/or their relatives

Year	Event	Response
1985	Bombing of Terminal 1	No CISM activities; support by the Medical Services unit
1986	Bombing of the Mail Distribution Center at the Airport	No CISM activities; support by the Medical Services unit
1986	Bombing of the US Airbase	No CISM activities; support by the Medical Services unit
1986	Suicide attempt	Psychological support

Year	Event	Response
1988	Pan American Airways bombing (Lockerbie, Scotland)	Immediate counseling (help) by the airport clinic for flight attendants and station employees 1:1 activity (help) by the director of Medical Services
1996	Birgen Air Accident near the Dominican Republic	Counseling (help) for relatives
1997	Bombing of a bus with German tourists in Cairo, Egypt	Counseling (help) for less seriously injured tourists during their travel home
1997	Terror attack on tourists in Luxor, Egypt	Support for the relatives after repatriation of the corpses of the murdered German (Mental Health P. and Peers)
1997	Traffic accident of tourist bus in South Africa (death of a tourist and injuries of accompanying relatives)	Counseling (help) for the returning tourists and meeting their relatives (Mental Health P. and Peers)
1998	Murder of German foreign aid worker in Kenia	Counseling (help) for relatives upon return of his wife (Mental Health P. and Peers)
2000	Concorde crash in Paris (Air France)	Counseling (help) for relatives of crash victims before their flight to the memorial service in Paris. Assistance for employees of the affected airline (Mental Health P. and Peers)
2000	Attack on tourist camp in Sri Lanka	Counseling of returning traumatized victims and relatives picking them up at the airport (Mental Health P. and Peers)
2001	World Trade Center Attacks	Activities (help) for United Airlines flight attendants 1:1 counseling

2003	Death on duty (accident)	1:1 counseling (help) for employees and debriefing of the medical emergency crew
2004	Death of an employee on his way home from work	1:1 counseling (help) and defusing
2004	Death of the managing supervisor on duty in the operations center	1:1 counseling (help) and defusing
2004/ 2005	Tsunami in Asia	1:1 counseling Peer support
2005	Suicide of FRAPORT employee (jumped off a building, 14th floor)	1:1 counseling (help) and defusing

The official team in charge of counseling and help was formed in 1996 and consists of doctors, psychologists and employees of the security unit at Frankfurt Airport.

This team conducted all response missions with different setups and is in charge of the concept for the development of a larger team.

Structure of CISM at FRAPORT

In addition to the core team (12 team members) in 2005 there is a group of approximately 140 trained employees (peers) which are available for critical events.

The members are primarily responsible for taking care of meeters and greeters. This team will be basically supplemented by employees of the internal and external pastors. The taking care of helpers, others involved in the mission, airline (crew, staff, check-in, terminal) will be done by the CISM team of Medical Services. If required, this team will be supported by CISM pastor team members and ATC-AP Team.

Training measures will continue. The counseling team is to include a total of 280 persons.

FRAPORT AG has already close cooperation with the airports in:

- Hahn, Germany
- Hannover, Germany
- Saarbrücken, Germany
- Leipzig, Germany (partnership)
- Dresden, Germany (partnership)
- Cairo, Egypt (management since 2005)
- Lima, Peru (management since 2002)
- Antalya, Turkey (management since 2003).

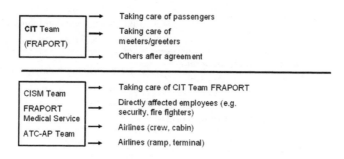

Figure 9.3 **Responsibilities of the FRAPORT Crisis Intervention Team (CIT), CISM Team, medical services, and the joint Air Traffic Control – Airport (ATC-AP) Team**

Many airlines such as:

- Delta Air Lines (USA)
- United Airlines (USA)
- British Airways
- Ethiopian Airlines,
- Air France already used the experts of the Medical Services at Frankfurt Airport whenever there was a need of medical and (!) CISM support.

This international group can react quickly and is well qualified based on the number of languages spoken.

Questions about the training program have already been received from national (Leipzig) and international airports (Antalya).

This kind of service will be always provided in close cooperation with our partner DFS/ATC. Figure 9.4 illustrates the system of the ATC-AP Team, where two different companies will cooperate with a corporate team leadership.

By operating an international airport there are many different organizations such as Federal police, police, customs, management etc. which in general have different interests that have to be coordinated.

At Frankfurt Airport we set up the ERIC (Emergency Response and Information Center) unit where all the managers meet in one of the above-mentioned cases.

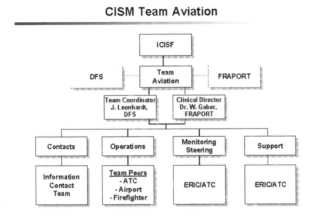

Figure 9.4 Structure of the CISM Team Aviation (Air Traffic Control – ATC-AP)

Rules and Procedures

Considering many past missions with a different setup of the involved teams and varying challenges, there is the necessity to have clear structures and rules. The reason for this is that team members have different personalities, handle stress in different ways and have varying qualifications. Now they come together as a team (maybe for the first time) and must function as experts when facing a large number of affected persons. Similar to the work of pilots, it turns out that it is good to work with checklists in order to ensure uniform and qualified assistance for all affected persons. This gives team members more security if they can rely on checklists, notes and other *instructions*. This is something very normal and in no way means that the helper does not have the situation under control.

Nobody would accuse a pilot of not doing his job well because he is using checklists.

Lessons Learnt

- Use existing procedures
- Create trust
- Know your partner
- Regular communication (3 to 4 times a day)
- Make agreements/decisions
- Keep the agreements (no independent actions of an individual)
- The co-operation with external partners must be optimized.

Advantage and Benefits

An international airport is subject to a huge number of incidents and influences based on many different situations. These are not always reasonable or understandable for an outsider. However, an intervention or cut in the chain of procedures calls for a tremendous and long-time development and coordination process to in the end guide the airport with its 52,000,000 passengers without major problems.

This means continuous information and involvement of external organizations that bring themselves into the procedures and structures based on clear guidelines to ensure smooth operation of the airport.

FRAPORT AG, as owner and manager of Frankfurt Airport, is well prepared for different scenarios up to a disaster due to experience and international consulting activities.

The procedures and operating instructions for emergencies are a cornerstone for cooperation with external partners (health department, city fire department, Federal police, customs, airlines etc.).

For logistic reasons the airlines must often rely on competent assistance in order to react to not planned situations/events and to ensure immediate and extensive counseling for affected persons in addition to all further special measures that are required.

Chapter 10

Case Study Lake Constance (Überlingen)

Jörg Leonhardt, Carol Minder, Sabine Zimmermann, Ralf Mersmann and Ralf Schultze

Summary

This chapter outlines the debriefing as Critical Incident Stress Management (CISM) intervention on the basis of a tragic accident. Firstly, the method is described theoretically. However, a special focus is laid upon the description of three different perspectives:

- On the first level, the perspective of the mental health professional MHP, who was responsible for the team, is taken. The single steps of the debriefing and the coherent considerations are outlined.
- On the second level, the perspective of the peers – as team members during the debriefing – is focused.
- On the third level, light is shed on the perspective of those, who experienced a critical incident and participated in the debriefing.

It is our concern to go beyond a merely theoretical description and to offer the reader a practical-oriented and living approach of the debriefing.

The Debriefing

Critical Incident Stress Debriefing CISD was the first crisis intervention technique for groups developed within CISM (see Chapter 3). The older literature even used CISD synonymously for CISM. By now, CISM has developed into a comprising method with different tools, which are based on each other and also intertwine. Therefore, CISD is one special method of CISM among others. The Critical Incident Stress Debriefing is a group based intervention tool comprising seven phases (see Chapters 3 and 5). The debriefing requires guidance by a mental health professional and peer support.

Generally, the debriefing takes two to three hours depending on group size and intensity of the session. The debriefing is neither a therapy session, nor does it substitute a therapy. Debriefing is an acute crisis intervention to help overcoming Critical Incident Stress Reactions.

The group should be homogenously in the sense of all participants being equally exposed to or involved in the incident. Hierarchy is of no importance. Also, the event should have passed already.

The necessity of a debriefing is indicated by the symptoms concerning the whole group, behavioral changes, existing symptoms or increasing severity of symptoms, the intensity of the critical incident, and the subsequent impairment of the people affected. An extensive description of the debriefing is given by Mitchell and Everly (2001).

The Accident

CISM Debriefings after the Mid-air Collision over Lake Constance

Background On 1 July 2002 at approximately 23:50 hours there was a mid-air collision of two aircraft north of the town Überlingen near Lake Constance in Southern Germany. The aircraft involved were a Tupolew TU 154 M en route from Moscow (Russia) to Barcelona (Spain) and a Boeing 757-200 en route from Bergamo (Italy) to Brussels (Belgium).

Both aircraft were being controlled by the area control center in Zurich (Switzerland) at the time of the collision. The accident happened over German territory.

A total of 71 people were on board of both aircraft. No passenger survived; on the ground were no casualties.

CISM at DFS In 1997, the DFS began to set up a CISM program based on the International Critical Incident Stress Foundation (ICISF) standards. DFS today has 80 trained peers. All peers have successfully completed the following courses:

- Individual Crisis Intervention and Peer support
- Basic Group Crisis Intervention
- Advanced Group Crisis Intervention.

All DFS peers have been certified by the International Critical Incident Stress Foundation, ICISF (see Chapter 7).

At the time of the accident, the Swiss air navigation services organization skyguide had its own Crisis Management Program. However, it was not possible to use this program given the fact that the whole organization was affected by the accident.

The DFS Crisis Management Board met during the night of the 1st of July to assess the situation. In the morning of the 2nd of July, DFS offered to support the Swiss colleagues with the DFS Crisis Intervention Team.

Four days later skyguide accepted the offer and called for the assistance of the DFS Crisis Intervention Team.

Preparing the Team The DFS crisis intervention team leader started to recruit the team according to the following criteria:

- Peers must volunteer for involvement
- Peers must work in a radar control center
- Male and female peers are needed
- Peers must have experience in CISM

The team consisted of two team leaders (Mental Health Professionals, MHP) and five peers. The team was given all the necessary information. They communicated via e-mail and telephone and met at Frankfurt Airport to depart to Zurich.

The team outlined the following requirements for skyguide:

- Up to 20 people per group
- Roughly equal exposure to the accident
- Participants may not work in operations after the debriefings
- Sufficient time should be scheduled for debriefings (at least 3 hours)
- Drinks should be provided after debriefings
- Talks should be held with the management before and after debriefings.

Debriefings in Zurich After their arrival in Zurich, the team was updated. A meeting was held with the management of the Swiss air navigation services, during which the management learned about CISM and was advised that the contents of debriefing sessions are confidential and cannot be disclosed.

Based on the list of participants provided by skyguide, the team endeavored to decide on optimal and homogenous groups for the debriefings.

Since the air traffic controller working the night of the accident was Danish, nationality was also a criterion for establishing the groups. Other criteria used to determine the grouping included how close the controllers were to the event and to the controller involved.

All of the controllers were able to highly identify with the event, the controllers affected and the organization. The more the controllers identified with the situation, the more traumatized they were potentially.

The accident did not make an impression on any of the controllers' sense of hearing, sight or smell.

A total of six debriefings were held with a total number of 120 participating controllers. Four debriefings took place six days after the accident, and two debriefings took place ten days after the accident. A hotline was set up for the next six months. A follow-up session took place five months after the accident.

The Personal Initial Situation of the Peers On 2 July 2002, we received a call from the DFS crisis intervention team asking for the participation of two UAC Karlsruhe peers to support the Zurich air traffic controllers after the catastrophe of Überlingen. The collision had happened the day before and we, the five Karlsruhe peers, prepared for a CISM intervention at the DFS radar center *Rhein-Radar*. A few discussions had taken place the day before with the colleagues, who had worked the last nightshift and who therefore had to follow the collision on the radar screen. Interventions for more colleagues were planned. As the air space of the Karlsruhe Center borders the Zurich air space, all colleagues were emotionally involved very much. This was also true for the air traffic controllers in Langen, as they are bordering Switzerland in the lower air space. As a result, two controllers (male) from Karlsruhe and one controller (female) from Langen were recruited for the CISM team to support Zurich Center. As the chosen controllers were so close to and familiar with the event, a high acceptance of the Swiss colleagues was expected. Due to the adjacent air spaces, the controllers worked closely together in their daily routines. This aspect is not to be neglected, as the Zurich colleagues did not have any CISM experiences until then.

The anticipation to support more or less affected persons in larger groups caused mixed feelings, as we got acquainted with debriefings in theory only. Our experiences with CISM were limited to conversations with one or two persons after an infringement of separation. Consequently, we questioned and feared, if we could handle these special demands. We did not want to make any mistake and the incident was very drastic for us as well. Also, we highly identified with the Swiss colleagues. However, the challenge to help outweighed the doubts and therefore we intensively prepared the debriefings according to what we knew from the courses.

Preparations Before departing for Zurich the DFS team leader provided us with all the necessary information on the planned process of the debriefings, the structure of the groups with the participants' position, e.g. controller, technician, air traffic service assistants, and administrators.

Arriving in Zurich, we made ourselves familiar with the premises. A large room was necessary for the group meetings with enough chairs, which we set up in a circle. Add to that, the supply with drinks was important. We, as the CISM team, needed a separate room for meetings and pauses.

Before each debriefing, the roles and positions for each team member were fixed. We agreed upon an open communication within the team, e.g. no use of secret signs such as hand signs etc. was allowed.

A 3D Perspective to the CISM Debriefing

The Debriefing follows a seven step model (Chapter 3 and 5): The debriefing commences on a cognitive level, prosecutes step by step the emotional level and afterwards enters the cognitive level again at the end of the process (Figure 10.1).

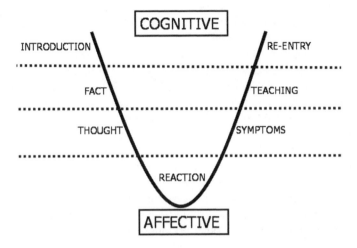

Figure 10.1 Seven phases of debriefing (modified according to Mitchell and Everly 2002)

Introduction Phase

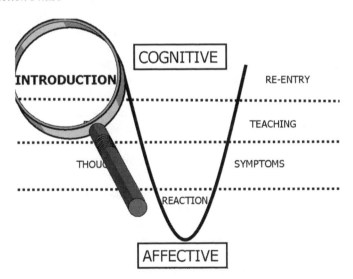

Figure 10.2 Introduction phase

Theoretical Considerations The CISM team members introduce themselves and explain the rules and procedures of the debriefings. It is emphasized that the debriefing does not depict an investigation. Everybody speaks for himself during the debriefing and it is stressed that there is no room to criticize organizations or persons.

The role allocation of the CISM team and the single team members were introduced as well as their positions.

It is pointed out that nobody is forced to say anything, but at the same time that any comments might be helpful to others and therefore will be an important input.

During this phase, it is aimed at making the procedure transparent, diminishing fear and tension, and at stressing that debriefing is not equal to psychotherapy. The presence of the peers highlights the latter fact as well.

The introductory phase creates the initial position to start the debriefing procedure from. The persons affected are regarded as equal partners; it is communicated openly within the group as well as in the team itself. This optimizes the faith in the team, the process, and the own coping strategies. The avoidance of criticism leads to the participants' concentration on themselves and their own reaction. Criticism and anger are understandable; however, they distract from one's own reactions and provoke a projection onto something else or other persons.

View of the Peers The team members introduced themselves and roughly explained their job position. The procedure and the aim of the debriefing were made transparent. As CISM was completely new to the Swiss colleagues, it was important to give deep insight into the background and aims of CISM debriefings in order to free the participants from skepticism and fear. As peers we knew, the controllers would expect honesty and frankness. That was why it was indispensable to highlight the open communication within both the team and the group in the introductory phase already. Also, in this phase it was decisive to characterize the debriefings as not being part of an investigation or psychotherapy. Moreover, the statement was crucial that criticism, anger, and disappointment about the management and media were understandable, but not very helpful as these factors only distract from own reactions. Our general attitude and position were meant to appreciate the controllers affected as equal partners and as focus of the debriefings.

View of the Participants Our employer recommended all operational staff to participate in one of the debriefings organized and offered by the DFS CISM group. Since there was no further explanation, the participants did not know what to expect and what the term debriefing really meant.

After a brief introduction, the procedure was quite clear and the fact that the peers were ATCOs from Germany made communication fairly easy. They were familiar with our working procedures, so we did not need to explain every aviation related term. Some of the debriefing groups were quite large, but that did not pose any problems since everyone knew each other from work.

Fact Phase

Theoretical Considerations As in all the other phases of the debriefing, the team leader commences the fact phase with a question. The members of the group takes turn in answering this question. The fact phase serves as platform for a personal introduction of each participant and their job position as well as for a description of their view of the incident.

In the special case of the debriefing after the catastrophe of Überlingen, the facts were quite clear and everybody was conscious of them. Besides the introduction of the participants, we had to think of an appropriate fact phase considering the special circumstances. We could not neglect the fact that the participants of the debriefings have not been directly involved in the accident.

The question we posed was the following: *We would like you to tell us your name and your job position. Also, we are interested in when and how you have heard about the accident.*

All participants are updated. A common ground to start from is created. This does not refer to the facts of the incident itself, but to the facts concerning how it was perceived and which information has been gathered by the different persons. The first and most striking information on the incident manifests itself in the time and the way this particular information is perceived. In most cases, intensive feelings stay linked to this situation. The first impressions depict an avenue to the reactions related to the incident. Untreated however, these impressions might trigger the same reactions in the future over again.

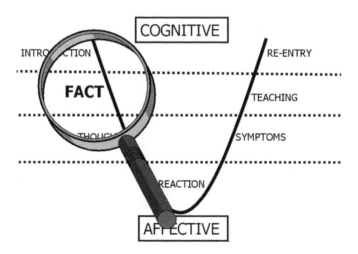

Figure 10.3 Fact phase

In contrast to debriefings with rescue workers, where the creation of a coherent picture is focused, in this case the common denominator is the point of time and the way, the persons affected found out about the incident.

The day the accident entered their lives.

View of the Peers During this phase it is important that all participants have their say in order to build up a common ground to start the debriefings from. In the debriefings in Zurich, the point of time and the circumstances, how everyone got to know about the incident, depicted the decisive factor. This situation was common to all, had burned into their minds and therefore served as initial situation for the debriefing.

Already in the introductory phase, but also in the fact phase some participants released their emotions. That made it difficult for the peers to stick to the debriefing structure. Yet, the team leader helped to follow the structure. The clear role allocation of team leader and us as peers gave us support to concentrate on our tasks. The role assignment of the different peers made it possible to give our undivided attention to the individuals.

View of the Participants The first question was relatively easy to answer, as we could talk about facts and made a good start into the debriefing.

In the fact phase the different ways people had heard about the accident became evident and the concern for the others was roused. We realized that everyone was affected, even people who had not shown any reactions in the days before the debriefing. We learned that people reacted in different ways and that they expressed their thoughts differently as well.

Generally, the memories of the situation, in which one had heard of Lake Constance for the first time, were still powerful. Therefore, they were linked to intense impressions and thoughts.

We appreciated that the DFS CISM team members as external persons, not belonging to our circle of friends, showed genuine concern. During the time following the accident, we as controllers felt accused by the media and the general public. Most people did not understand why we felt accused and it was therefore helpful that some friends and the DFS peers had a complete comprehension of our emotions.

Thought Phase

Theoretical Considerations The thought phase focuses on the first cognitive level, the embedding of the experience and the associated emotions and reactions. The first thoughts associated with the incident might be unconsciously memorized. The emotional memory is hereby supplied. All further reactions, triggered by the emotional memory, are fixed by the unconscious memorization of these first thoughts. By verbalizing the first thoughts in this phase of the debriefing the probability, that similar reactions occur, is minimized. Even if they occur, the intensity of the reactions is reduced.

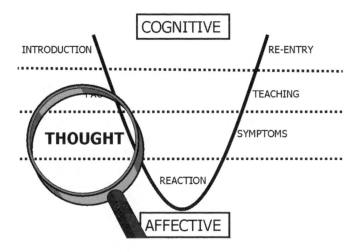

Figure 10.4 Thought phase

The participants again take turns answering the team leader's question: *Which was the first thought that struck your mind and which thought have you been unable to get rid of since?*

The thoughts are a link between the cognitive and the emotional area. Thoughts can evoke an emotional reaction. If thoughts are verbalized, they can be reflected upon and as a consequence they loose their emotional strength. The verbalization of thoughts, externalizing them, enables the person to cut the bond with the emotional memory. By doing so the bond becomes apparent and tangible.

If thoughts are not verbalized, but are still a severe strain, a vicious circle can develop with a pull that makes it difficult to break out of the circle.

The homogeneity of the group facilitates the expression of thoughts: the presence of the controller on duty in that particular night would have certainly restricted the free expression of thoughts for example.

View of the Peers If possible, everybody should get his or her say in this phase of the debriefing as well. It is important to become aware of the first thoughts after the critical incident. The thoughts depict the bridge between cognitions and emotions.

The homogeneity of the group was the decisive factor – especially in this phase – for the willingness to participate in the discussion. The presence of managers, for example, or of the controller on duty at the night of the incident probably would have stopped many participants from expressing their thoughts freely. Rational persons, which air traffic controllers definitely can be characterized as, get access to their emotions more easily via their thoughts and cognitions. In this phase, we as peers became aware of the way in which the emotionality of the incident approaches us.

As we are controllers ourselves, we felt an increasing emotional tension. Leaving the cognitive level is scary at first, as you do not know were your emotions might lead to. The ability to understand this fear made it easier for us as peers to perceive and bear the statements and the emotionality of the participants.

View of the Participants   Most of the answers were something like "What if this happened to me?", "How can I continue working?", "How was this possible, how can something like this happen in Europe?", "What if I had been on duty that night?" and "What is going to happen to the controller who was working that night?".

The participants were mainly concerned about the controller on duty and about the thought that something similar could happen to them.

Because this question was very concrete, it was still rather easy to answer, although you were talking about personal fears in front of a group of people. When it became obvious that others had similar thoughts and concerns, the uniqueness of your own feelings gave way to a sense of connectedness and emotional closeness with the other group members.

Today, as trained peers, we know that the thought phase is very important. Talking with family members or friends about such an event does not replace a debriefing, because they never have the same thoughts as the colleagues, who are affected by the situation the same as you are.

Reaction Phase

Theoretical Considerations   The reaction phase depicts the most emotional phase of the debriefing process. In this phase the group is posed a question, but the participants do not take turn to answer like in the other phases. Whoever feels able to answer does so.

What was the worst for you personally? Imagine the story of the accident as a movie; you have the possibility to cut out the worst part of this movie. From the time you knew about the incident until now, which part would you cut out?

The questions of the reaction phase aim at the most serious and emotionally strongest aspect. If the participants can name and perceive the most straining and most striking part, they probably can handle the less straining aspects themselves as these parts are less threatening. Dealing with the most straining aspects activates the coping process.

The participants' worst experience is emotionally very intense. However it is perceived as less threatening in case it is voiced in a protected group situation and under the guidance of professional counselors, who control the process.

The metaphor of the movie is an offer for a mental coping model. It is not to be misunderstood as support to repress or even belittle the experiences. The metaphor constitutes a link to the coping process.

By externalizing and by imagining the events as a movie sequence the disaster turns tangible and treatable. The incident looses its power, because it is named and finds its way out. The incident loses its threat, because the persons affected gain

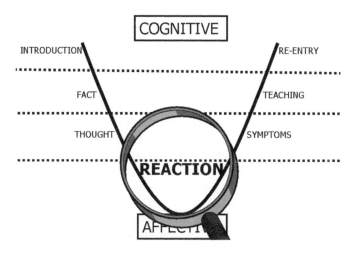

Figure 10.5 Reaction phase

their control back again as they are able to cope with the incident mentally. Images, which you can cut out, will be less threatening in the dreams or are even expelled from the dreams. The metaphor that you are able to do something with the movie, to be able to examine it, to cut it, to put it aside und to get a grip on it, gives the person control again. The person cannot change what has happened, but he or she can learn to handle the emotions. The metaphor helps to regulate the continuum of closeness and distance: How close do I want the emotions to be and what degree of closeness can I bear?

The coping process starts immediately.

During this phase, it is optional to make a statement. Experience shows that even participants, who do not actively take part in the discussion, comprehend the process internally and start their personal coping.

Internally comprehending the metaphor – which might also happen automatically while reading this chapter – launches an active coping process.

The homogeneity of the group is important to avoid traumatizing others with even worse images.

View of the Peers The persons affected were asked to become aware of those aspects concerning the incident, which had the most serious and emotional strongest effect on them. During this phase the participants had the option to make a statement or to just listen. However, we pointed out that every contribution would be beneficial to others to internally comprehend the process and start the coping phase.

At this point of the debriefing, we as peers were affected by the emotionality of the group and our own emotions as well. The structure of the debriefing and the profound training as peer helped to emotionally support the group. It was important to us, to show our own emotions without being overwhelmed by them. We as peers have to hold the balance between showing empathy and being affected ourselves.

View of the Participants Some of the answers, which came up, were "The worst part was to imagine his face when it happened", "Not knowing who had been on duty", "Why him?", "Why us?", "The certainty that it was a mid-air collision".

When asked to describe which aspect of the accident was most difficult to cope with, some people did not know how to express their feelings. With the help of the film sequence you want to cut out it became easier.

This was the most emotional part of the debriefing. Some people were crying, others refused to answer the question.

It also became clear that those participants who had a closer relationship to the controllers involved in the accident were more affected, because they were extremely worried about their future. This shows the importance of the homogeneity of the group.

Symptom Phase

Theoretical Considerations The symptom phase builds the bridge to return from the emotional to the cognitive level. The symptom phase plays a decisive role for the transfer of the whole group from the level of strong emotionality to cognitive control.

The symptom phase focuses on possible changes that the participant observed during the preceding phases. Moreover, a link to critical incident stress reactions is established. The symptom phase serves as an evaluation of changes that have already taken effect. These changes might justify a one-to-one session or a referral to medical or psychotherapeutic aid.

The question was: *Have you noticed any changes in your behavior, thinking or any physical changes?*

A variety of changes is possible. The necessity of referrals has to be checked. Sometime some examples for possible changes need to be mentioned. This can either be done by the peers or the group members.

By building the bridge, namely the description of the emotions or symptoms, the person affected is enabled to assess his or her emotional reactions cognitively and therefore his or her individual coping strategies are activated. The person affected is led back to his individual coping process and can fall back on coping competencies, which had been paralyzed by the intensity of the emotions.

Taking this step of the process means the person affected gains control over the reactions in the sense of an independent coping. The fear to react in an uncontrolled manner and to be incapable of handling these reactions is minimized. The own reactions, emotions, and symptoms become normal when comparing them with the

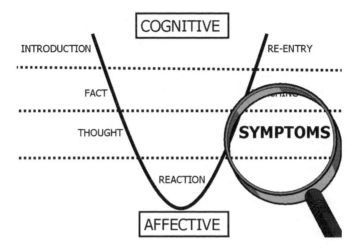

Figure 10.6 Symptom phase

others of the group: the participants realize that they are not unique and not on their own, but that others feel and react the same or very similar.

The feeling of uniqueness disappears and gives way to the solidarity within the group and the similarity of feelings, symptoms and reactions. Without the structured and protected process the debriefing offers such symptoms would not be talked about and the chance for coping would be minimized.

View of the Peers The symptom phase led the participants back from the emotional to the cognitive level. The CISM team together with the participants tried to find out by examining the descriptions of the mental and physical state and the respective symptoms, whether specific symptoms became fixed or chronic or whether some changes could be detected. The need of further support or psychotherapeutic treatment was assessed.

The persons affected tried to grasp the emotions on the cognitive level by describing their reactions, which is already a part of the active coping process. In a way, the reaction can be controlled instead of just re-acting to the fear. To talk about feelings by taking a step aside and observing oneself from the outside is an appropriate way for control-oriented persons, which air traffic controllers definitely belong to.

This phase was very demanding on our concentration and attention: on the one hand, we had to see to it that a further measure was commenced if the description became too intense; on the other hand, we had to memorize the described changes in order to refer to them in the following phase.

As peers we were able to make some authentic contributions in this phase, as we described our own feelings and the ones of our DFS colleagues' right after the catastrophe. We felt great solidarity with the Swiss colleagues and expressed this feeling, which surely encouraged some participants to make their statements.

View of the Participants Examples of answers to this question were "I've lost confidence in my working methods", "I sleep badly", "I have no appetite at all", "I'm avoiding TV news and big newspaper headlines".

Almost everybody had noticed changes in either his behavior or his working methods. Controllers found it difficult to do their daily job without confidence in their working tools. During normal shift work we did not have the time to talk about our thoughts and concerns, so the debriefing created the possibility to do that. The realization that everyone was affected and had trouble coping with daily life made it easier to work together again.

Teaching Phase

Theoretical Considerations The teaching phase depicts the active phase of the peers. The team leader commences the phase, but then leaves it up to the peers to report from their experiences, to lecture on stress and stress management and finally to summarize the comments of the group. When summarizing the contributions of the participants it is important to refer to reality and the situation of the group members. Ready-made concepts and advice are of little help. It is rather the described problems and changes of the participants that should be effectively used.

There is no guiding question asked in this phase.

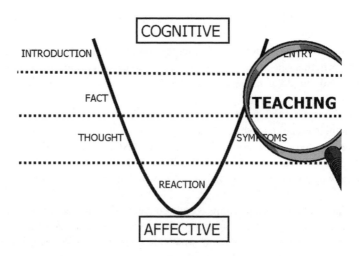

Figure 10.7 Teaching phase

The peers sum up the comments and try to minimize the feeling of uniqueness and to normalize the symptoms. The normalization of the reaction means in this context to label the reactions and irritations of the persons affected as understandable and comprehensible. The normalization aims at the reaction and not at the incident itself.

The main issue in this phase is the demonstration of ways to cope with and handle the critical incident stress and how participants can reactivate own coping strategies.

When demonstrating ways of concrete, pragmatic, and active reaction management it makes sense to pick up, sum up, and integrate the group's comments.

View of the Peers This is our active phase as peers: we tried to sum up what had so far been said and to highlight the common ground. Thus, we diminished the uniqueness of the reactions and normalized them. We stressed that the incident was so radical and drastic for everybody and that diverse irritation were to be expected and normal.

When demonstrating ways how to deal with the critical incident stress, we tied up the existing coping strategies.

We tried to encourage the participants to actively and pragmatically deal with the reactions, for example by thinking of concrete measures to support the coping process. As peers and air traffic controllers we were able to offer adequate support as we reported how we managed the situation ourselves and gave examples of our behavior at work or our daily routines at home.

The advice was quite simple and concrete, yet many times persons in an extremely stressing situation cannot recall them automatically.

Some suggestions were: to talk to colleagues; to accept that you cannot perform 100 per cent in your job; to demand support and support from the supervisors and support them in return; to plan your daily routine and to stick to it to avoid finding yourself in a hole; to do sports and to carry on with your hobbies; to go out with your colleagues or with a group; to be open for family and friends and to tell them frankly if you need your space.

It is important to see to it that you do not expect too much of yourself and to consider the reduced capacity to perform.

We had the impression this particular phase was very important, because we could clearly show the ones, who were highly strained due to the incident, that they were still in control of their life.

View of the Participants The teaching phase was helpful, because we received concrete suggestions and options for action. The German peers also talked about personal experiences, which added credibility and made it easier for us to accept their advice.

Re-entry Phase

Theoretical Considerations The re-entry phase is the last step of the debriefing process. By now, everybody should have reached the cognitive level again. You can tell by the faces, the postures and the comments that the group has gone through the process together. Also, you could tell how soothing it has been for everybody to talk about the emotional status and how important this has been even if threatening in the beginning.

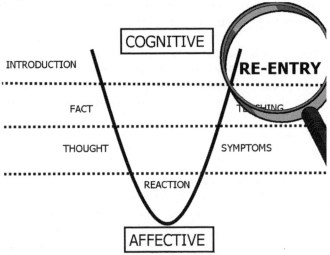

Figure 10.8 Re-entry phase

Realizing to be able to deal with the situation and realizing that this takes time puts the participants back into control and turns their focus on the future at the same time.

The team leader commences the re-entry phase again with a question: *Despite all the bad that has happened, do you find any positive aspect or something you could take as a chance?*

This question has to be introduced very carefully, as it can easily be misunderstood or even be rejected. It is not the point to redefine the incident as something positive; it is rather the aim to extract the positive aspects of the own reactions and to figure out the *lessons learnt*.

Mentally, the view is focused on the future during this phase by planning the next steps concretely: Which are my aims, plans and perspectives? Which are the exact steps I will take to reach these aims?

The question enforces a constructive future perspective. The head is mentally raised and the view is turned forward. The whole posture is changed.

The participants are enabled to find their own way out and their feeling of control is strengthened.

No aspects are repressed. It is pointed out that the reactions can last for a while or can even reoccur, usually with a weaker intensity, however.

View of the Peers We tried to outline a perspective for the immediate future. As peers, it was our major concern to highlight once again that it would take time to perform 100 per cent again. Also, we stressed the importance of looking at the own working methods self critically and honestly. Repeating orders for the pilots or asking them to repeat something automatically comes along with the situation. This behavior rather proves the sense of responsibility than a deficit or lessened professionalism as controller. Acting like this facilitates the stepwise approach towards complete productivity. Nobody calls such working methods as unprofessional.

We thanked those involved for the frankness and willingness to talk about very personal and often painful feelings. At the end of the debriefing, we did not forget to offer the possibility of single counseling with the team members and to offer a contact number for further questions.

After the actual debriefing, most of the participant stayed in the room, had a drink and took the chance to exchange some more thoughts. Similar to our Swiss colleagues we experienced this as important and comforting at the same time, because this gathering deepened the solidarity.

View of the Participants It was very difficult to find any positive aspects, and some people found it outright impossible. Nobody had any idea in which ways this accident would affect our company and our working environment. We also could not imagine how the controller carrying this burden would ever be able to live a normal life again.

Some points, which were mentioned, are: "I will say NO in the future and not let others push me into difficult situations", "I will start working again carefully and reconsider my own working methods".

The re-entry phase differed somehow between the groups. In the first debriefing, held only one week after the accident, events were still very recent. Some of the participants of later debriefings already found a way to look forward to the future. It also depended on the duration and the emotional intensity of the debriefing. However, it is absolutely necessary to focus on the future, and it can even be helpful to just listen to the discussion without actively participating in it.

Conclusion from the Participants

We experienced how important and valuable mutual international support can be. When confronted with such a big accident, the identification within the organization is too strong, making it impossible for the own CISM team to help. Of course there are a number of general crisis intervention teams in every country. In such a specialized environment however, getting support from people with the same profession makes everything a great deal easier.

In the light of different languages and cultures, such support can only be successful if the CISM measures taken comply with established standards.

Main Topic in the Debriefings: Identification

As expected the other controllers were in a situation to strongly identify their self with the colleagues who are directly involved in the accident, the organization and profession as air traffic controllers. For the Danish controllers working in Switzerland, nationality was a predominant theme. Identification was the overriding issue throughout all of the debriefings. Identification and the related emotions were so strong and dominant that the controllers could not identify with the victims. We can assume that if they had also identified with the victims, the emotional effects would have been so overwhelming that it would have been too difficult to deal with the reactions.

Sympathy with the relatives of the victims was expressed in all debriefings in the teaching and re-entry phases.

Conclusions

Despite the fact that the Swiss air navigation service provider skyguide already had a Crisis Management Program, it still took a lot of time to explain to the management how CISM and the debriefing sessions work. As a result, not all of our guidelines were observed, e.g. some air traffic controllers should have gone to work after the debriefing. We were able to convince the managers not to do so – after a considerable amount of time had passed.

It is advisable in such situations and crises to have a contact person at the place of the accident that is trained in CISM and can take the necessary precautions. It is also helpful if the organization receiving the support runs a CISM program and therefore speaks the *same language* in crisis intervention.

Consequences

In 2003, skyguide began to set up its own CISM program in Zurich and to train peers. Skyguide follows the standards established by the ICISF and has become a close cooperation partner of DFS.

DFS is calling for CISM to be introduced as a standard by air navigation service providers in Europe. CISM should be set up and introduced in line with ICISF standards; the peers should be trained and certified by a recognized coach.

In cooperation with DFS the International Federation of Air Traffic Controller Associations (IFATCA) and EUROCONTROL organized a congress on CISM in

ATC, which was held in Bucharest in October 2004 (see also network described in Chapter 7).

DFS is the project leader in an international consortium writing the CISM Implementation Guidelines for EUROCONTROL.

Reference

Mitchell, J. and Everly, G.S. (2001), *Critical Incident Stress Debriefing: An Operations Manual for CISD, Defusing and Other Group Crisis Intervention Services* (3rd ed., Ellicott City, MD, USA).

Chapter 11

Training in Critical Incident Stress Management – Principles and Standards

Victor Welzant and Amie Kolos

Introduction

It can take only a second for a traumatic event to change a person's life forever. In an era where sudden and unpredictable crisis and traumatic events are likely realities for approximately 60 per cent of men and 50 per cent of women (Kessler, Sonnega, Bromet, Hughes and Nelson 1995), survivors may benefit from immediate supportive intervention. During the past two decades, a number of crisis intervention programs have evolved, both individually and in conjunction with hospitals and centers (Roberts 2005). These programs, which include 24-hour telephone hotlines and specialized intervention teams, require a strong commitment and extensive training for those providing the intervention. One such intervention, which emerged in the 1980s, Critical Incident Stress Management (CISM), has been a major proponent of peer-based crisis intervention.

The terrorist attacks on 11 September 2001, the tsunamis that destroyed Asian islands in December 2004, and even more recently, Hurricane Katrina, which demolished New Orleans, Louisiana in September 2005, yielded an increased need for acute crisis programs. Following traumatic events since the 1980s, those in occupations that were at high risk for exposure to trauma saw that properly administered CISM was instrumental in resolving some survivors' early signs of distress, and many became interested in helping with this intervention process. While the desire to help those in crisis is a good starting point, it is crucial that those providing intervention are properly trained. This includes training in and the practice of such basic crisis skills as empathic reflection, active listening, in addition to displaying hopefulness in order to help the person in crisis realize that their situation is not hopeless (Roberts 2005).

At the field of Critical Incident Stress Management evolved, it became clear that preparation and training is a key to the development of competence (Everly and Langlieb 2003). Everly (2002) outlined five core competencies in emergency/ disaster mental health. These core competencies include:

1. The ability to differentiate benign vs. malignant psychological symptomatology

(psychological and behavioral triage)
2. Skill in one-on-one crisis intervention
3. Skill in small group crisis intervention (20 or less)
4. Skill in large group crisis intervention (20-200)
5. The ability to plan and implement an integrated, phasic multi-component emergency mental health initiative.

A number of studies have concluded training that focuses on one or more of these core competencies enhance effectiveness (Nunno, Holden and Leidy 2003; Pearce and Snortum 1983; Tierney 1994). In one study that tested the effectiveness of training in suicide intervention, participants were significantly more competent in assessment of risk factors, and direct questioning about suicidal ideation (Tierney 1994).

Just as each specialized crisis intervention model has its own methods for training, there are courses designed specifically for training interveners in CISM. These courses are regularly updated and evaluated. A growing number of empirical studies are demonstrating the benefits of CISM, which include processing the traumatic event, assembling resources and social support, and reducing signs of stress and trauma (Robinson 2000).

Training in Critical Incident Stress Management

The strength of any crisis intervention training program lies in its ability to reliably develop and enhance skill sets for crisis interventionists. The International Critical Incident Stress Foundation has created standardized training curriculum and provides instruction for trainers interested in teaching CISM courses. When training is standardized, practitioners have a common set of principles and guidelines by which intervention may be provided. This standardization provides needed structure, which serves as an antidote to crisis (Everly 2006).

As noted above, Everly (2002) has outlined core competencies for the provision of a comprehensive, multicomponent crisis intervention system. The training sequence of courses, outlined by the International Critical Incident Stress Foundation (2005), addresses these requisite skills through *core curriculum* which consists of four workshops, ranging from 14 to 16 hours each. The following sequence of training highlights the workshops and briefly outlines topics included in the training.

ICISF Core Curriculum

One must learn four core components of crisis intervention in order to be effective in a crisis situation. These four courses are: (1) Assisting Individuals in Crisis, (2) Group Crisis Intervention, (3) Strategic Response to Crisis, and (4) Suicide: Prevention, Intervention, and Postvention. Each of these courses provides the trainee with a background and skills to address commonly encountered situations in the field of crisis intervention.

The first core course, Assisting Individuals in Crisis, is an introductory training program that emphasizes an overview of crisis intervention principles, Critical Incident Stress Management core components, active listening and responding skills, psychological triage in crisis situations, and a model of crisis intervention for use with individuals.

Group Crisis Intervention, focuses on assisting the participant in developing skills to work with those in crisis in groups and providing four models of group intervention. These models, in addition to guidelines and contraindications for their use, are based on empirically derived crisis intervention principles. Practice scenarios allow participants to develop skills in group facilitation.

In Strategic Response to Crisis, the third of the core courses, trainees practice the skills for both assessing the needs of those in crisis and developing a comprehensive, phasic, and multicomponent response to complex crisis situations. These skills are taught and emphasized through progressively more difficult scenarios during the course of the training.

The last core course, Suicide: Prevention, Intervention, and Postvention, is a skills based training designed to provide trainees with the knowledge and strategies to recognize, prevent and respond to suicidal individuals. This course also has an emphasis on referral skills and resources. Training people to work with individuals in groups following a suicide in an agency is also a component of this training.

Taken together, these core trainings offer a sequenced opportunity to develop skills for effective intervention. It is recommended that the core curriculum be followed by specialty training, in the specialty area in which the practitioner will be serving those in crisis. As noted below, the combination and sequence of training is designed to progressively build skills incrementally, with the goal of mastery. It should also be noted that training in Critical Incident Stress Management is based on a model of peer support that utilizes those from similar occupations to provide support with proper mental health practitioner oversight in as part of an intervention team.

Advanced and Specialty Training

In addition to the core courses, it is advised that all CISM practitioners complete the Advanced Group Crisis Intervention Course. This training program has been designed to provide participants with the latest information on critical incident stress management techniques and post-trauma syndromes. The program emphasizes a broadening of the knowledge base concerning critical incident stress interventions as well as Post Traumatic Stress Disorder, which was established in the CISM: Group Crisis program, and/or in publications. Specifically, the CISM: Advanced Group program addresses advanced concepts which are the foundations of CISM and, more specifically, the group interventions of defusings and Critical Incident

Stress Debriefings (CISD). In CISD and CISM special emphasis is placed upon troubleshooting difficult and complex situations. Further information about the courses, including training topics for the core courses, can be found in Appendix A.

The training sequence outlined above prepares participants for providing Critical Incident Stress Management services in a variety of situations and circumstances. As with any human service field, the evolution of new applications of CISM has led to the creation of new training programs as well as efforts to provide a systematic sequence of training for those interested in specialty areas of CISM application. The ICISF has developed the Certificate of Specialized Training (COST) Program to address the requests from practitioners for this recognition of the differences in applying CISM in various professions. Currently, there are six specialty areas in which one can obtain a COST certificate (ICISF 2005). Each area of specialty requires the core curriculum as outlined above, and in addition 2 additional specialty training (13-14 hours each) as well as an elective training (13-14 hours). The specialty areas currently are:

- Mass Disasters and Terrorism Specialty
- Emergency Services Specialty
- Substance Abuse Crisis Response Specialty
- Workplace and Industrial Applications Specialty
- Schools and Children Crisis Response Specialty
- Spiritual Care in Crisis Intervention Specialty.

As CISM training has evolved, practitioners have found that the specialization training has enhanced their ability to address more specific concerns in their occupation with the various groups of those in crisis that they serve. Practitioners in the airline and other transportation professionals have found that the Workplace and Industrial applications specialty as well as the Mass Disasters and Terrorism Specialty are especially relevant to their practice. Interested practitioners may refer to the ICISF Certificate of Specialized Training (2005) or to Appendix B of this chapter for further discussion of the training curriculum.

CISM Training Principles

The training of crisis intervention practitioners must continually seek to improve and integrate the latest evidence based practice standards while remaining practical and useful under field conditions. The following principles may serve as a guide to the training of CISM teams:

1. Training in CISM is most effective when a sequenced program of skills is taught and practiced. As outlined above, ICISF recommends a sequence of core training prior to advanced and specialty training ... even for experienced professionals. This is designed to assure standardization in training and

practice.
2. Peer based crisis intervention is the core of CISM. It is believed that trained peers in various occupations (for example emergency services, air transportation industry) add credibility and allow for effective provision of occupationally based crisis intervention services following traumatic events in the workplace.
3. CISM training should follow adult learning principles (Kalafat 1984). This requires respect for the experiences that the participants bring to the training as well as an interactive and participatory training style.
4. Training should be updated and continuing education is necessary to maintain and enhance competence, as well as to allow interventionists access to the latest evidenced based and *lessons learned* strategies and tactics for crisis intervention and management.
5. CISM training is an integrated process that BEGINS with the core training components outlined above. This is the first step in developing mastery; however this must be followed with careful supervision, and further team training using adult education principles. Potter, Stevens, and LaBerteaux (2003) have masterfully discussed the use of training scenarios for crisis team training and continuing education. Post incident critiques can prove particularly effective in promoting learning if done respectfully and confidentially. Additionally, clinical oversight by trained and experienced mental health professionals is essential for integrating training into effective integration. Mentoring programs that provide support for CISM/Crisis team members extend the peer support concept to the development of team members skills and confidence as well.
6. Effective practice includes adherence to standards of care and training (Everly and Mitchell 1999). Practitioners must remain familiar with current standards and changes in CISM and other forms of crisis intervention. Continuing education, literature reviews, and collaboration with other practitioners may all serve as methods to ensure adherence to current standards. Training in CISM should only be done by those who have completed a trainer's curriculum through ICISF. CISM trainers use recognized course materials as the foundation for their training to ensure standardization.

Becoming an ICISF Approved Instructor

To effectively standardize the training of Critical Incident Stress Management, ICISF has created an approved instructors program. This program invites experienced advanced practitioners of crisis intervention to attend an additional curriculum to teach ICISF courses. Instructors are provided a standardized course manual, instructors' materials and provided with background material and teaching insights in a 20 to 24 hour training program for each course they desire to teach. Each instructor registers their training courses with ICISF and course evaluations are reviewed by ICISF as part of a quality assurance initiative. Additional information about becoming a trainer is available from ICISF.

Conclusion

Critical Incident Stress Management offers a model of intervention for workplace trauma that is particularly effective in high risk environments for exposure to traumatic events. The airline industry is one example of such a venue. Peer based CISM services merge the principles of crisis intervention and resiliency with the value of those who know well the demands of an occupation through first hand experience. With proper training guidelines, interventionists will have the needed skills available to them when crises occur, thus allowing for effective service provision. As CISM continues to evolve, evidence based and practical experience will allow for enhanced training curricula, reflecting the best that science and practice have to offer the field of early psychological intervention following trauma.

References

Everly, Jr., G.S. (2006), *Assisting Individuals in Crisis, Fourth Edition.* (Ellicott City, MD: International Critical Stress Foundation, Inc.).

Everly, Jr., G.S. (2002), Thoughts on training guidelines in emergency mental health and crisis intervention. *International Journal of Emergency Mental Health* 4:3, 139-142.

Everly, Jr., G.S. and Langlieb, A. (2003), The evolving nature of disaster mental health. *International Journal of Emergency Mental Health* 5:3, 113-118.

Everly, Jr., G.S. and Mitchell, J.T. (1999), *Critical Incident Stress Debriefing: An Operations Manual for CISD, Defusing, and Other Group Crisis Intervention Services, Third Edition.* (Ellicott City, MD: Chevron Publishing Corporation).

International Critical Incident Stress Foundation (ICISF) (2005), *Certificate of Specialized Training.* (Ellicott City, MD: International Critical Incident Stress Foundation, Inc.).

Kalafat, J. (1984), Training community psychologists for crisis intervention. *American Journal of Community Psychology* 12:2, 241-251.

Kessler, R.C., Sonnega, A., Bromet, E., Hughes, M. and Nelson, C.B. (1995), Posttraumatic stress disorder in the National Comorbidity Survey. *Archives of General Psychiatry* 52, 1048-1060.

Nunno, M.A., Holden, M.J. and Leidy, B. (2003), Evaluating and monitoring the impact of a crisis intervention system on a residential child care facility. *Children and Youth Services Review* 25:4, 295-315.

Pearce, J.B., and Snortum, J.R. (1983), Police effectiveness in handling disturbance calls. *Criminal Justice and Behavior* 10:1, 71-92.

Potter, D., Stevens, J.A. and LaBerteaux, P. (2003), *Practical Concepts and Training Exercises for Crisis Intervention Teams.* (Ellicott City, MD: Chevron Publishing

Corporation).

Roberts, A.R. (2005), Bridging the past and present to the future of crisis intervention and crisis management. In A.R. Roberts (Ed.), *Crisis Intervention Handbook: Assessment, Treatment, and Research, Third Edition* (pp. 3-34). (New York: Oxford University Press, Inc.).

Robinson, R. (2000), Debriefing with emergency services: Critical incident stress management. In B. Raphael and J.P. Wilson (Eds.), *Psychological Debriefing: Theory, Practice, and Evidence* (pp. 91-107). (New York: Cambridge University Press).

Tierney, R.J. (1994), Suicide intervention training evaluation: A preliminary report. *Crisis* 15:2, 69-76.

Chapter 12

Cost Benefit Analysis of a Critical Incident Stress Management Program

Joachim Vogt and Stefan Pennig

The preceding chapters described the application of CISM in aviation. The benefits of the program are very often obvious for all operative staff and make it a must in modern aviation. However, not every aviation organization has CISM implemented by now. One reason might be that managers who have to decide on the implementation of a CISM program want to avoid the financial and organizational effort or do not see convincing evidence of benefit. This chapter is therefore dedicated to provide arguments that CISM is by far worth the effort due to strategic and also economic benefits. The aims of this chapter are:

- to show the strategic importance of CISM for aviation organizations
- to document monetary as well as intangible benefits of CISM
- to offer theoretical and methodological inspiration for future cost-benefit-analyses of CISM programs.

In an effort to assess and document the business benefits of a CISM program, a EUROCONTROL feasibility study on the applicability of economic evaluation methods to human resource programs was conducted in 2003. The combination of human resource evaluation (Kirkpatrick 1994; Phillips 1996) and utility analysis (Boudreau and Ramstad 2003) was the theoretical basis on which the feasibility study instruments were developed. On the basis of the experience with the feasibility study, a new framework for cost-benefit-analyses was developed: The Human Resources Performance Model (HPM). HPM is the basis for the main study which is currently conducted on behalf of EUROCONTROL and the German Federal Institute of Occupational Safety and Health.

In the following, the two evaluation models will be presented as well as the core results of the two studies.

Feasibility Study

The theoretical model underlying the economic evaluation method used in the feasibility study is the Human Resources Evaluation Model (Kirkpatrick 1994). It was originally conceived to classify and determine the effects of staff trainings. The original model consists of four levels:

1. Reaction: Describes the satisfaction of participants with the training measure (content, method, organizational framework, behavior of the trainer, etc.). This level was obtained by asking CISM peers to evaluate their CISM courses and asking the controllers about their reactions to consulting a peer.
2. Learning: Refers to the learning effects of the training, whether or not they are practically applied (behavior). The peers were for example asked about the knowledge, abilities, and attitudes they obtained during the CISM courses, and the controllers reported what they learned from consulting a peer.
3. Behavior: Comprises behavioral changes on the job indicating the transfer and adoption of learning effects in the work situation. In the feasibility study, the peers were asked to what extent the CISM qualification prepared them for their task and in what respect it changed their behavior during a consultation. The controllers were asked to what extent consulting a peer helped them to recover and return to work.
4. Results: This level obtains the resulting effects of behavioral changes as relevant for the organization. These include:

    - product quality
    - efficiency of performance
    - client satisfaction
    - image
    - staff satisfaction, fluctuation, and absenteeism
    - cost reduction.

In the feasibility study, the impairment of essential air traffic controller (ATCO) skills after the critical incident was assessed to determine what factors changed in demand characteristics in the person-task interaction. Moreover, the ATCOs self-assessed their performance after a critical incident (CI) and how CISM supported the recovery of their work ability.

Later, another level to the model was added (Phillips 1996), which in the framework of the feasibility study was labeled:

5. Impact: The profitability of the training as return on investment (ROI), which is measured in monetary units.

    The full cost of training is estimated together with the turn-off (ROI = [revenue − costs] / costs). In the feasibility study, the return on investment of CISM was measured as the estimated cost reduction by gained controller days due to accelerated return to work relative to the full costs of the program per controller. This final step is the core of utility analysis (Boudreau and Ramstad 2003), while the previous steps belong to the human resource evaluation.

The four and five level models were applied to a double chain: Reaction, learning, behavior of the peers after their training were considered as well as reaction, learning,

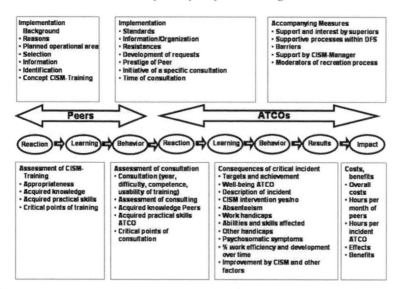

Figure 12.1 The evaluation model of the feasibility study

behavior, results, and impact on the part of the ATCOs after a CISM intervention (see Figure 12.1).

The boxes in Figure 12.1 show the different data that were obtained in the feasibility study. Due to the pilot character of the investigation, only 61 people were involved as subjects:

- The manager of the CISM program, who was mainly asked about the implementation background and process (two top left boxes in Figure 12.1) as well as the costs of the program (bottom right box in Figure 12.1).
- 13 peers, who assessed the training they received and the consultation they gave thereafter (two bottom left boxes in Figure 12.1); they also reported about the organizational support for their work (top right box in Figure 12.1).
- 47 ATCOs, who experienced a critical incident and gave information on the consequences (penultimate bottom right box in Figure 12.1); among these were 20 ATCOs who requested CISM support after a CI and could estimate the contribution of CISM to their recovery.

A program can only be evaluated against the objectives it was implemented for. Therefore, the CISM-Manager, the peers and the ATCOs were asked to report the originally intended goals of the CISM program. The ATCOs primarily mentioned the enlargement of communication possibilities after CI (supervision), whereas the peers emphasized the avoidance of long-term emotional consequences, which in the case of Post Traumatic Stress Disorder (PTSD) could lead to substantial loss of performance capability. The achievement of these objectives was estimated to be very successful.

The ATCOs who requested a CISM consultation after the CI were asked how long it took to recover to full working ability and how much CISM contributed to this recovery. On average they reported that CISM accounted for 36 per cent of their recovery. This corresponded to a reduction of the time off by one and three days after a CI and the Überlingen accident (see Chapter 10), respectively. Relative to the full costs of the program in the investigated sample this accorded to a conservative approximation of 250 per cent return on investment. The exact calculation is published in Vogt, Leonhardt, Köper and Pennig (2004).

Besides the accelerated recovery which can be expressed in monetary values, the controllers and peers mentioned additional benefits. The overall benefit of CISM for the organization and the personal benefits for the peers and ATCOs in the feasibility study were estimated extremely high, even by those ATCOs who after CI did not consult a peer. Some of them reported that CISM is a back-up for them, which can be relied on in situations which are too difficult to cope with alone. From the peers' perspective, CISM created a better culture of error handling, trust, and more accessibility. These documented CISM-induced cultural improvements thus can be assumed to have strategic importance to aviation organizations because safe transport is the core business.

Main Study

Theoretical Background

On the basis of the experiences with the evaluation model of the feasibility study, a new concept for steering and evaluating human factors (HF), human resources (HR) and training (T) programs was developed. The Human Resources Performance Model (HPM) uses the strengths of five managerial evaluation/steering models and integrates them in a new, comprehensive framework. HPM facilitates the advantages of the following five models:

1. The *Human Resource Evaluation*, partly used in the feasibility study, structures the psychological functional chain of competence-inducing interventions.
2. The *Utility Analysis*, partly used in the feasibility study, translates behavioral effects into economic effects and at the same time provides decision aids in the planning of HF/HR/T interventions.
3. The *Human Capital Bridge Model* structures the planning and effect levels of interventions and combines entrepreneurial aims, strategy relevant human capital and single interventions. Especially the selection of behavioral and performance related indicators, which are also employed in utility analyses,

are systematically related to the success-inducing tackling of bottle-necks.
4. The *Performance Improvement Model* extends the perspective by systematic consideration of enterprise performance. The effectiveness and cost-efficiency of interventions thus is not considered isolated but in interaction with other behavioral, structural or technological bottle-necks and according interventions.
5. The *Balanced Scorecard* (Kaplan and Norton 1996) helps planning, evaluating and steering an organization on the basis of controlling indicators. It is a long-term strategic tool.

The HPM is introduced as a framework that pursues the claim:

1. To permit an overall evaluation of HF/HR/T management of a company.
2. To illustrate economic and behavioral effects of single HF/HR/T interventions.
3. To connect the evaluation of HF/HR/T interventions with the overall view of the company's human capital.
4. To provide a basis for decisions on investments in the HF/HR/T management.
5. To identify causal connections between input and output in HF/HR/T management to allow optimization and adaptation.
6. To illustrate interdependencies between HF/HR/T investments and other, for example technical investments, in order to highlight synergetic effects and bottle-neck factors within the complex system.

With respect to the CISM program of air navigation service providers, the following effects were hypothesized on the basis of the feasibility study (see Figure 12.2):

1. Arrow 1: a CI, for example an unexpected separation infringement between two aircraft, causes psychological (and maybe also physical) reactions of controllers with insufficient own coping capabilities (see also battery metaphor in Chapter 4).
2. Arrow 2: on the individual level, these reactions alter abilities (for example slower reaction time), on-the-job-behavior (for example more generous separation) and performance (for example less capacity) of the controller; the alterations have to be measured against the personal baseline (reaction time, separation habit, capacity before CI).
3. Arrow 3: the impairments on the individual level have an impact on the operational management by supervisors and chief of sections (for example controller is planned out); air traffic control (ATC) process quality and efficiency are affected; on the strategic level, the organizational goals of

safety, punctuality, and productivity are in danger.
4. Arrow 3 represents a significant step; up to now HF experts focus on the individual and maybe group level; however a systematic combination of HF data with for example financial controlling does not take place; we have to do this significant step – cross the psycho-economic Rubicon.
5. Arrow 4 indicates that a cost-benefit analysis of CISM must prove that the reactions to the CI are appropriately treated and thereby prevent negative effects on abilities, on-the-job behavior, and performance.
6. Arrow 5 indicates that CISM also has a direct organizational impact; in the feasibility study, for example, benefits with respect to communication culture were reported by ATCOs who never ever had a CISM consultation.

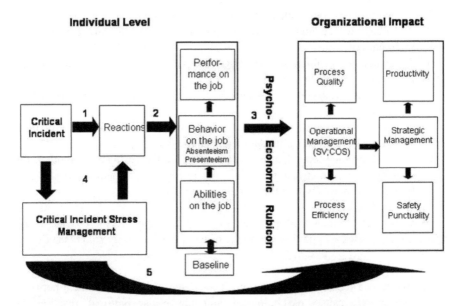

Figure 12.2 Model of expected benefits of CISM in an air navigation service provider (ANSP; SV: supervisor; COS: chief of section)

Objectives of the Main Study

The main study followed five main objectives:

A Enlarge the database for efficiency and profitability conclusions on CISM:
   - Larger and more representative sample of ATCOs than in the feasibility study
   - CIs limited to separation losses during the last two years
   - Online monitoring of performance recovery after CI.

B Specify the cost and benefit effects:
- What are the real cost reductions, monetary benefits, and goal-attainment contributions of CISM.

C Analyze the causal chain:
- Investigate the processes of cost reductions, monetary benefits, and goal-attainment contribution
- Test the hypotheses of CI induced performance losses and CISM induced individual and organizational improvements which were deducted from the feasibility study.

D Study the influence of operational leadership (supervisors, chiefs of sections):
- Investigate further the impression that group leaders play an important role in dealing with CIs, responding to impacts on ATCO performance, and state-of-the-art use of CISM.

E Investigate further the interaction of CISM, communication, and safety culture:
- Follow the hint in the feasibility study that CISM supports a constructive handling of CIs and thereby improves organizational culture in general.

Participants and Procedures

In the period June to October 2005 1,030 questionnaires were delivered to the five largest DFS radar centers and the six largest DFS air traffic control towers (with the exception of Frankfurt Tower; Frankfurt Approach was included instead). The questionnaires covered the following issues:

1. Basic self assessment regardless of CI (baseline):
   - Abilities for example selective attention, stress resistance.
   - On-the-job-behavior for example monitoring and planning of traffic.
   - Performance for example average number of aircraft under control.

2. Description of the most serious CI during the last two years and its effects:
   - Relative to baseline abilities, on-the-job-behavior, and performance.
   - Emotional and psychosomatic effects.
   - Potential effects on safety, errors, work flow, capacity, cooperation with assistant, adjacent sectors, staff planning.

3. Assessment of communication culture and changes due to CISM implementation and practice.

4. Long-term health effects of untreated CI.

Of the 1,030 questionnaires only 30 per cent were returned, i.e. 309. Of these 77 contained descriptions and effects of CIs during the past two years. The performance in the week before (baseline) and the week of the CI was rated as well as in the six weeks after the CI. Another ten ATCOs self-reported a CI during the six months of data gathering; they monitored their psychological state and work performance not on a weekly but on a daily basis. Among the 87 ATCOs with a past or current CI were 51 who consulted a peer.

A detailed analysis of the 87 CIs and their effects revealed that 21 did not match the definition of CI or Critical Incident Stress Reactions (see Chapter 4). The results reported here were therefore obtained on the basis of the remaining 66 questionnaires, among them 48 reporting a CISM intervention.

The sample is substantial and representative for the total number and quality of CI which have occurred in the German airspace during the years 2004 and 2005. Objective A can therefore be assessed as fully met.

Results

According to the Human Resources Performance Model (HPM), three levels have to be considered:

- The self-management of the ATCO (individual level).
- The decentralized process management by supervisors and chiefs of sections (process level).
- The central, strategic management (strategic level).

Costs and benefits of CISM occur on all three levels and depend on their interaction. The causal chain starts with the coping of the ATCO after a CI, continues with the supervisors' response (for example plan the ATCO out, limit traffic, call a peer), and ends with the impact on the performance and staff management of the organization.

The Handling of CIs on the Individual Level

Figure 12.3 shows a modal of causal relationships of CIs and CISM on the individual level. Moreover, the interfaces to the process level (top), the pilots and the team (left hand) are indicated.

The data on this level revealed which abilities, on-the-job behavior, and performance aspects were altered after the CI (ATCO performance management in Figure 12.3). Moreover, it shows which coping mechanisms the ATCO could rely on and which could be developed with the support of CISM.

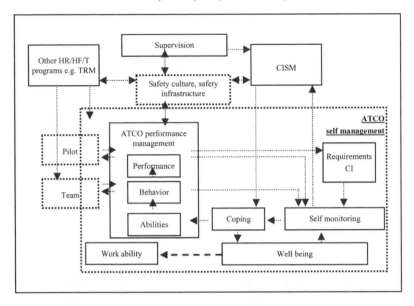

Figure 12.3 Self-management of air traffic controllers as process model (individual level: HR: human resource; HF: human factor; T: training; TRM: team resource management)

Performance Curve and Presenteeism

The ATCOs with CI during the last two years rated their performance in the week before the CI, in the week the CI happened, and in the following six weeks (weeks 1–6). Figure 12.4 shows the performance curve of the ATCOs. In contrast to the previously reported results of the feasibility study, ATCOs who participated in CISM show a similar recovery curve than their colleagues who chose not to consult a peer. This is partly due to the selection of standard CIs for the main study, while the feasibility study contained also the Überlingen accident and events that did not match the definition of a CI (see Chapter 4).

With respect to work-related benefits of CISM, the presenteeism area in Figure 12.4 is of great interest. In the week the CI happened, the ATCOs reported on average a reduction of their performance by 20 to 25 per cent. However, only two ATCOs had days off (absenteeism). In all other cases, ATCOs worked with a minor performance deficit, which is called *ill at work* or presenteeism. The presenteeism area was equally large for controllers with and without CISM. It will be shown later that the average curves hide several CISM and non-CISM sub-groups and that the immediate consultation of CISM peers after a CI makes a big (positive) difference. The presenteeism area would be much larger if the ATCOs would not have had the possibility of CISM.

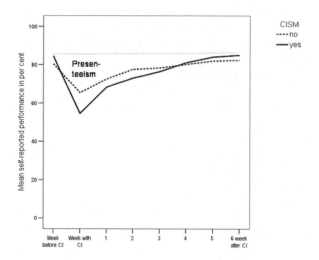

Figure 12.4 Self-reported performance in the week before, during, and after a critical incident (CI) or controllers requesting or rejecting CISM

The presenteeism effect can be expressed in lost controller days if we integrate the area above the curves up to the normal performance which was – in accordance with many workload studies – reported to be 80 per cent. Over the seven weeks, the presenteeism cumulated to 5.4 lost days per ATCO having a CI.

The recovery curve is steepest in the first two weeks after the CI. The ATCOs reported that over 50 per cent of their recovery is facilitated by the communication with peers and other colleagues.

Quality and Quantity of Impairments

Especially the ATCOs who did not receive CISM treatment reported significant impairments of over 20 per cent in fundamental cognitive abilities like for example planning, mental flexibility, vigilance, selective attention, and multitasking. Also social skills were affected like emotional stability, or the ability to cooperate.

The ability reductions correspond to the reported changes in on-the-job behavior: Complex data intake and processing on the one hand and team work and cooperation on the other were impaired especially in the group of controllers who had no CISM after a CI.

As a consequence, the ATCOs reported noticeable effects on the quality of the ATC process. Impairments of smooth traffic flow, traffic capacity, and cooperation (with colleagues, assistants, adjacent sectors, or pilots) were reported. Again, ATCOs without CISM reported more performance losses (30 per cent) than ATCOs with CISM (15–25 per cent).

ATCOs with CISM reported ability losses over 20 per cent only for two items: stress resistance and emotional stability. This indicates that they realized the impact of the CI on their personal coping capability in stressful situations, however, their self-reported on-the-job behavior and professional performance changed below 25 per cent. Their performance in normal operations seems to be less affected than in the group of ATCOs who did not participate in the CISM program.

Facilitating CISM Effects

In the answers to the open question how CISM supports their coping, the ATCOs mentioned that they appreciated talking the CI and their reactions over with a peer. They needed to follow the CI up and to clear its emergence. The importance of normalization (see Chapter 4) and structuring everyday life were frequently mentioned as helpful. We can conclude that CISM supports the primary self-coping mechanisms of the ATCOs.

The Importance of Operative Leadership (Process Level)

Figure 12.5 shows a model of causal relationships on the process level with the interfaces to the individual level (bottom) and the strategic level (top). How do supervisors handle CIs and potential performance losses thereafter? What initiatives are taken by supervisors to support the ATCO, the coping process, and organizational learning about the CI? How do these initiatives affect individual and organizational performance with respect to the strategic goals?

The ATCOs with CIs reported the response of their supervisors to the CI and potential performance loss. The responses could be classified in six categories:

1. Immediately planned out, CISM consultation and return to work thereafter (22 cases).
2. Immediately planned out, CISM consultation and end of duty for the current work day (CISM no ops, 6 cases).
3. Immediately planned out, CISM consultation and further consultations during the following days (several CISM, 7 cases).
4. No supervisor intervention, delayed CISM consultation (delayed CISM, 13 cases).
5. No supervisor intervention, no CISM, but support by colleagues or supervisor (6 cases).
6. No supervisor intervention, no CISM, no support (12 cases).

The average emotional upset of the ATCOs was equally large in all groups (82 per cent). The supervisor interventions thus were fairly independent of the emotional upset of the ATCOs.

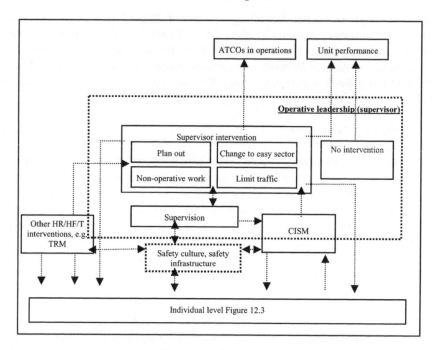

Figure 12.5 Operative leadership as a process model (process level; HR: human resource; HF: human factor; T: training; TRM team resource management)

Effects of Supervisor Interventions

The best performance recovery was found in group 2, the ATCOs who were immediately taken off duty, consulted a CISM peer shortly after, and were given time off for the rest of their work day. In comparison with the other groups, it became apparent that the performance recovery did not only depend on the supervisor intervention but also on the immediate state-of-the-art application of CISM (subsequently called standard CISM in contrast to delayed CISM). Group five for example, although reporting social support from colleagues or supervisors, experienced more emotional and psychosomatic symptoms like sleeping problems, irritability, feelings of guilt etc. than the groups receiving standard CISM.

The group differences in the performance recovery curves support the finding of the feasibility study (Vogt, Leonhardt, Köper and Pennig 2004) that ATCOs not participating in CISM cannot be considered a control group. The comparison of CISM-ATCOs with non-CISM-ATCOs can be misleading because the four CISM-groups (groups 1–4) and the two non-CISM-groups (groups 5 and 6) were very heterogeneous in themselves. For example:

1. CISM group 1, the ATCOs who received standard CISM but had to finish their shift, reported on average the second strongest performance loss in the week of the CI and the second flattest performance recovery curve (baseline performance was reached six weeks after the CI).
2. CISM group 2 reported on average the smallest performance loss in the week of the CI and the shortest recovery time (back to baseline performance in the second week after the CI). Considering that all groups experienced equal emotional upset after the CI this indicates that the immediate replacement of an ATCO after CI in combination with standard CISM treatment and giving the ATCO time off for the rest of the shift is best practice.
3. CISM group 3 (several CISM) reported on average the strongest performance loss in the week of the CI and the longest time to return to baseline (six weeks after the CI still 5 per cent impairment). The ATCOs in this group consulted a peer several times. Their initial performance recovery curve (first week after CI) is nearly as steep as in group 2.
4. CISM group 4 (delayed CISM) experienced on average a moderate performance loss in the week of the CI. Their recovery curve resembled that of group 5 indicating that delayed CISM supports recovery not significantly better than other social support. Baseline performance was reached five weeks after the CI.
5. Non-CISM group 5 (social support without CISM) experienced on average a moderate performance loss in the week of the CI and a moderate recovery curve reaching baseline performance six weeks after the CI.
6. Non-CISM group 6 (no supervisor intervention, no CISM, no social support) reported the flattest recovery after a moderate performance loss. Like in group 5, baseline performance was reached in week six after the CI. However, the performance recovery curve in group 5 was steeper, especially during the two weeks after the CI and probably due to the experienced social support.

The picture of performance recovery is supported by the reported impairments of on-the-job behavior. Group 5 and 6 with no CISM treatment, reported on average 17–30 per cent impairment of these behaviors compared to only 5–10 per cent in groups 1–3. The experienced impairment in group 4, receiving CISM delayed, was 17–23 per cent. In order to avoid that CIs are coped with during work, standard CISM and time off for the rest of the work day after a CI are necessary.

Apart from their performance curve, the ATCOs were asked to report the impact on safety, work errors, traffic flow, and capacity. Group 2, ATCOs receiving immediate standard CISM and being planned out for the rest of their shift, reported no impairment here. Over 30 per cent capacity loss was described in the groups 5 and 6, receiving no CISM.

The results suggest that coordinated effort of ATCO self management, supervisor interaction, and standard CISM gives the best result for the individual and the organization. Best practice as ATCO is to ask for being released and to consult a peer after CI and as supervisor to call a peer for CISM, and to give the ATCO time off for the rest of the work day. The following chapter outlines the impact of best and poor practices on the strategic level.

The Impact on Strategic Leadership (Strategic Level)

In order to assess the strategic impact of CISM for the organization the above mentioned results have to be related to the strategic goals of the organization. The strategic goals of an air navigation service provider are safety, punctuality, and productivity. Where and how does CISM support these goals? To what extent are the costs for implementing and running CISM outweighed by the benefits? Figure 12.5 shows a model of causal relationships on this strategic level with the connection to operative leadership (see Figure 12.5) and individual performance management (see Figure 12.3) at the bottom as well as the airlines as costumers.

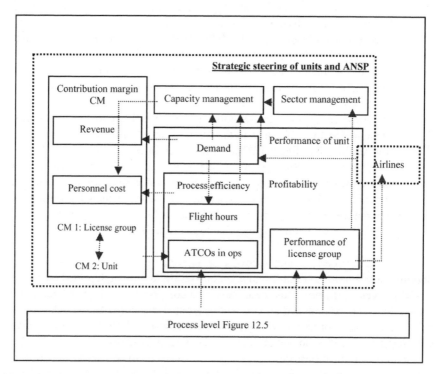

Figure 12.6 Strategic steering as process model (strategic level)

Return on Investment

Like in the feasibility study (Vogt, Leonhardt, Köper and Pennig 2004), which revealed a return on investment of about 250 per cent, a general profitability calculation was conducted in the main study. The total cost of the CISM program was 560,000 € since implementation in 1998. This figure contains training costs for peers, including the operative time lost while they participated in training. The total costs have to be related to the costs in the investigated sample of 309 ATCOs, which corresponds to 16.7 per cent of all German ATCOs, and the considered time period of 28 months (24 months of past CIs and four months of study conductance). This results in 560,000 € / 8 years * 0.167 * 2.33 years = 27,238 € for the 309 ATCOs and the 28 months on the cost side of the CISM program investigated in this chapter.

The benefit in the investigated sample can be calculated on the basis of the performance recovery curves and the reported contribution of CISM to the recovery in the following steps:

1. The recovery area in Figure 12.4 corresponds to 8.5 full work days per ATCO.
2. The ATCOs reported like in the feasibility study that 36.3 per cent of their recovery was due to CISM, which in combination with the 8.5 work days result in 8.5*0,363 = 3.0855 full work days per ATCO.
3. One ATCO work day costs on average 670 €.
4. The mathematical cost reduction due to CISM is therefore 3.0855*670 = 2,067 € per ATCO and CI.
5. 58 ATCOs in the investigated sample reported per definition CIs.
6. The total cost prevention is therefore 48*2,067 = 99,216 €.
7. The return on investment (ROI) is ROI = 99,216 / 27,238 * 100 = 364 %.

Like the feasibility study, the main study revealed apart from the ROI figure many intangible benefits with respect to the strategic goals of the organization. These are described in the following section.

Contribution of CISM to Strategic Goals

The three central strategic goals in ATC are safety, productivity, and punctuality. How CIs and CISM affect these goals was reported by the ATCOs in our sample:

1. Safety was not experienced as compromised after a CI, neither with nor without CISM. Although ATCOs reported ability impairments after a CI, they fulfilled their responsibility in regard to safety, worked more cautiously and thus compensated for CI induced performance losses.
2. Half of the ATCOs in group 6, not receiving any support, reported impairments of traffic flow during the first two weeks after the CI. Delays were explicitly mentioned in this context. In both groups without CISM support, every third ATCO worked with a reduced capacity after the CI.
3. The average capacity loss in the two non-CISM groups was about 10 per cent for 7.7 days. If we assume 100 annual CIs in a large air navigation service provider, which are not treated with CISM, the total productivity would be

impaired by approximately 0.02 per cent. This considerable capacity loss can be avoided by the best practice: planning the ATCO out, providing immediate standard CISM, and giving the ATCO time off for the rest of the work day.

The strategic goal of safety is supported by improvements in organizational culture especially with respect to communication and error handling. This was already proven in the feasibility study. The analysis of the 309 questionnaires of the main study confirmed this: The ATCOs rated on a five point change scale whether organizational culture described with an adjective list impaired or improved (-2 major impairment, -1 minor impairment, 0 no change, +1 minor improvement, +2 major improvement). With respect to all adjectives, organizational culture was reported to have been slightly improved by CISM. The attitudes *supportive* and *safety awareness* gained the highest average rating of +1.

Summary

Although the ROI calculations rely on some assumptions and estimations, they show – together with the descriptions of impairments in the ATCO questionnaires and the improvement of organizational culture – a strategic impact of CISM on the organizational goals of safety, productivity, and punctuality.

The results of the main study replicated the feasibility study results with an enhanced and representative data basis (goal A of the study). They illustrated the causal chain of CIs, individual performance loss of ATCOs, and subsequent impairments of traffic flow (goal C). Moreover, it could be documented, to what extent these effects were reduced by best-practice operative management (process level, goal D) and CISM. Costs and benefits were estimated as accurately as necessary for a general program profitability analysis (goal B). The ROI was estimated over 350 per cent and thus even higher than in the feasibility study. Also the positive impact of CISM on the strategic goals punctuality and productivity were demonstrated: total productivity without CISM would be impaired by approximately 0.02 per cent. With respect to safety, the main study replicated the feasibility study result that CISM improved organizational culture and also in this respect set heading to the positive (goal E).

In more detail, the main study results can be summarized as follows:

1. CIs impair individual ATCO performance over an average period of two weeks.
2. The performance of ATCOs receiving immediate and standard CISM recovers more quickly. With the help of CISM, significant ATCO impairments are prevented, which otherwise can occur especially with respect to selective attention, information intake and processing, traffic planning, communication and cooperation.
3. Without CISM, the individual impairments will affect the ATC process. This manifests especially in productivity (capacity) and punctuality. Per ATCO

with CI, CISM prevents about eight work days with an individual capacity loss of 10 per cent. Safety impairments due to individual performance losses are compensated by more cautious separation.
4. The return on investment within the CISM groups is over 350 per cent.
5. There are differences in the supervisor response and CISM application. Best practice is that the ATCO is released from duty, immediately receives standard CISM, and is given time off for the rest of the work day.
6. Social support cannot replace standard CISM. Although some ATCOs, who were not treated with CISM, reported very positive about support of supervisors and other colleagues than peers, they experienced more emotional responses and mental performance losses than ATCOs receiving standard CISM.

Recommendations

The following recommendations are given on the basis of these results:

1. The standard procedure for the handling CIs should be: The ATCO is immediately released from duty to get CISM peer consultation. The rest of the work day should be spent free or at least not in operations.
2. ATCOs, peers, and supervisors should be aware of the CI induced performance and process losses.
3. Operative leaders (supervisors, chiefs of sections) should be supported in their professional response to CI (see 1.), which facilitates the regeneration of performance on the individual level and maintains productivity and punctuality on the process and strategic level.
4. Peers should receive refresher courses on a regular basis and participate in annual peer conferences to support them in maintaining the good quality of their work.
5. The different human factors, human resources, and training initiatives in an organization should be coordinated and harmonized because they interact (Figure 12.3). Also the three levels of the Human Resource Performance Model, self management on the individual level, operative leadership on the process level, and strategic steering on the strategic level, should be linked.
6. The study reported here could be a benchmark for the integrative standardization, quality enhancement, evaluation, and development of human factors, human resources, and training initiatives.

Due to apparent overall positive effects of the program, efforts for providing CISM services and comparable human factor interventions should be supported throughout aviation. This was confirmed in discussions with operative and top managers of DFS (Chapter 2). Human factors are seen as key issues of the future; integration, harmonization, and standardization of these programs are therefore paramount.

Since there were differences within and between the CISM- and non-CISM-groups, the future strategy should be two-fold. First, in order to include the non-CISM-groups, the CISM intervention could be promoted as a tool for disseminating detailed information concerning the circumstances of an incident. Second, standard CISM in combination with operative leadership could be utilized to decrease the culture of self-blame that often surrounds CIs. Since many controllers interpret CIs, such as a separation loss, as a sign of failure and weakness, they are less likely to request a CISM intervention. They might be more likely to participate, however, if the program also included detailed information about the incident and how to prevent it in the future. Since CISM is associated with economic savings and strategic benefits, organizations should have an added incentive in enacting CISM programs.

References

Boudreau, J.W. and Ramstad, P.M. (2003), Strategic industrial and organizational psychology and the role of utility analysis models. In W.C. Borman, D.R. Ilgen and R.J. Klimoski (Eds.), *Handbook of Psychology: vol. 12. Industrial and organizational psychology* (pp. 193-221). (New York: John Wiley and Sons).

Kaplan, R.S. and Norton, D.P. (1996), *The Balanced Scorecard – translating strategy into action.* (Boston: Harvard Business School Press).

Kirkpatrick, D. (1994), *Evaluating training programs: The four levels.* (San Francisco: Barrett-Koehler).

Phillips, J.J. (1996, April), 'Measuring ROI: The 5th level of evaluation. Technical and Skills Training.', <http://astd.org/NR/rdonlyres/DOBCF259-880D-4EEC-BF89-7F1B9A88F430/0/phillips.pdf>, accessed 22 September 2004.

Vogt, J., Leonhardt, J. Köper, B. and Pennig, S. (2004), Economic evaluation of the Critical Incident Stress Management Program. *The International Journal of Emergency Mental Health* 6:4, 185-196.

Appendices

Appendix A: Core Curriculum Training Topics

This information, in addition to more information about each course can be found at http://www.icisf.org, ICISF's official website.

Assisting Individuals in Crisis

- The concept of CISM as a comprehensive crisis intervention program.
- The role of the individual crisis intervention in the comprehensive CISM program.
- Terms and concepts relevant to the study of crisis, traumatic stress and crisis intervention.
- Differential utilities of selected crisis communication techniques.
- Demonstration of selected crisis communication techniques.
- Psychological reactions to crisis and trauma.
- SAFER protocol for individual crisis intervention and it's role in comprehensive CISM.
- Demonstration of the us of SAFER protocol for individual crisis intervention.
- Review of common problems encountered while working with individuals in crisis.

Group Crisis Intervention

- Defining general, cumulative, critical incident stress and PTSD.
- High risk populations – emergency services, military, airline personnel.
- Defining Critical Incidents – death, injury, threat, terror, etc.
- Critical Incident Stress Management (CISM) fundamentals.
- Core intervention tactics.
- Demobilization.
- Defusing of small groups.
- Crisis Management Briefing.
- Defusing Demonstration.
- Essentials of group intervention, Critical Incident Stress Debriefing (CISD).
- Assessing the need for CISD.
- Factors which enhance CISM success.

Strategic Response to Crisis

- Core Concepts of early psychological intervention.
- Strategic Planning History and Background issues; Case examples.
- Incident Command System; Linking early psychological intervention plans to other agencies and organizations.
- Key Planning issues, concerns, and considerations.
- Planning exercises; Feedback and guidance.
- Planning for Mental Health Emergency Response Planning models; Comprehensive, integrated, systematic and multi-tactic approach to crisis management.
- Application of tactical components within a strategic plan.
- Final course exercise; Discussion/guidance.

Suicide: Prevention, Intervention, and Postvention

- Scope of the suicide problem.
- Statistical patterns.
- Common myths about suicide.
- Motivations for suicide.
- Attitudes about suicide.
- Small group role plays.
- Effective communication skills.
- Dealing with strong emotions.
- Developing an education briefing for suicide.
- Problem solving skills for crisis intervention.
- Small group role plays.
- Intervention skills for suicide prevention.
- Referrals.
- Postvention and aftercare.
- Care for the caregivers.

Advanced Group Crisis Intervention

- Nature of Human Stress.
- Stressors, Traumatic Stress.
- Symptoms of Excessive Stress.
- Stress-related Diseases.
- Stress Management.
- Recovering from Excessive Stress.
- Nature of Post-Traumatic Syndromes and PTSD.
- Early Warning Signs and Symptoms of Post-Traumatic Stress.
- PTSD versus Discipline Problem.
- Responding to Post-Traumatic Stress in Others – Management Reaction,

Peers' Reaction, Referral.
- A Rationale for CISD.
- Why CISD Works.
- Role of CISD in Prevention of PTSD.
- Current Thinking on Critical Incident Stress Management (CISM).
- CISD versus Psychotherapy.
- Supporting the Emergency Services Family.
- Expanded Role of Peer Support Personnel.
- Preserving CISM Team Members.
- Lessons Learned from CISM.
- Difficult Debriefings.
- Multiple Line of Duty Death.
- Symbolic Debriefings.
- Multiple Incident Debriefings.
- Comprehensive CISM approach to complex situations.
- CISM in Disasters.

Additional Courses

Please see the foundation's website for course outlines and objectives, as well as a schedule of conferences where each course will be offered.

- Assaulted Staff Action Program (ASAP): Coping with the Psychological Aftermath of Violence.
- CISM Application With Children.
- CISM Applications with Air Medical, Critical Care Transport and Airborne Law Enforcement.
- CISM in the Healthcare Setting.
- CISM: When Disaster Strikes.
- Compassion Fatigue.
- Corporate Crisis Response: CISM in the Workplace.
- Domestic Terrorism and Weapons of Mass Destruction: A CISM Perspective.
- Early Intervention and Crisis Response in EAP and Behavioral Healthcare Settings.
- Ethics For Traumatologists.
- Families and CISM: Developing A Comprehensive Program.
- From Trauma to Addictions.
- Grief Following Trauma.
- Law Enforcement Perspectives for CISM Enhancement.
- Line of Duty Death: Preparing the Best for the Worst.
- Pastoral Crisis Intervention.
- Pastoral Crisis Intervention II.
- Preventing Post-Traumatic Stress Disorder: Resilience Training to Build Psychological Strength.

- Preventing Youth Violence.
- Psychotraumatology For Clinicians.
- Responding to School Crises: A Multi-Component Crisis Intervention Approach.
- Stress Management for the Trauma Provider.
- TEAM: Team Evolution and Management.
- Terrorism: Psychological Impact and Implications.
- Thought Field Therapy.
- Treatment of Complex PTSD.
- Understanding Human Violence: A Primer for Emergency Services Personnel and Counselors.

Appendix B

Certificate of Specialized Training (COST) Program

Certificate	Mass Disasters & Terrorism Specialty	Emergency Services Specialty	Workplace and Industrial Applications Specialty	Schools and Children Crisis Response Specialty	Spiritual Care in Crisis Intervention Specialty	Substance Abuse Crisis Response Specialty
Core Curriculum Refer to 'Additional Information' for important info re: Core Curriculum These Three Courses Required	colspan across — *Complete four core courses as indicated below* Group Crisis Intervention and Individual Crisis Intervention and Peer Support and Suicide Prevention, Intervention and Postvention					
One Required			Advanced Group Crisis Intervention or Strategic Response to Crisis			
Specialty Curriculum Two completed courses are required	CISM: When Disaster Strikes	Line of Duty Death (LODD)	Corporate Crisis Response	CISM: Applications with Children	Pastoral Crisis Intervention	From Trauma to Addictions
	Terrorism: Psychological Impact and Interventions	Law Enforcement Perspectives	Understanding Human Violence	Responding to School Crises	Pastoral Crisis Intervention II	Stress Management for the Trauma Provider
	Domestic Terrorism	Families & CISM Stress Management for the Trauma Provider	CISM in the Healthcare Setting EAP/ Behavioral Healthcare		Grief Following Trauma	
Electives	One 2-day or Two 1-day	One 2-day or Two 1-day	One 2-day or Two 1-day	One 2-day or Two 1-day	One 2-day or Two 1-day	One 2-day or Two 1-day

Electives

2 Day Programs

- Assaulted Staff Action Program (ASAP): Coping with the Psychological Aftermath of Violence.
- CISM: Applications with Air Medical and Critical Care Transport.
- CISM: Application with Children.
- CISM in the Healthcare Setting.
- CISM: When Disaster Strikes: Linking Emergency Management with Stress Management.
- Compassion Fatigue.
- Corporate Crisis Response: CISM in the Workplace.
- Domestic Terrorism and Weapons of Mass Destruction: A CISM Perspective.
- Early Intervention and Crisis Response in EAP and Behavioral Healthcare Settings.
- Ethics for Traumatologists.
- Families and CISM: Developing a Comprehensive Program.
- From Trauma to Addictions (Critical Incidents, Trauma and Substance Abuse: Preventing Self Medication).
- Grief Following Trauma.
- Law Enforcement Perspectives for CISM Enhancement.
- Line of Duty Death: Preparing for the Best for the Worst.
- Pastoral Crisis Intervention.
- Pastoral Crisis Intervention II.
- Preventing Youth Violence.
- Responding to School Crises: An Integrated Multi-Component Crisis Intervention Approach.
- Stress Management for the Trauma Provider.
- TEAM: Team Evolution and Management.
- Terrorism: Psychological Impact and Implications.
- Understanding Human Violence: A Primer for Emergency Services Personnel and Counselors.

1 Day Programs

- Psychotraumatology For Clinicians.
- Treatment of Complex PTSD.

Please see http://www.icisf.org for additional information.

Index

Absenteeism, 161
Accident
   Eschede, train, 109
   Lake Constance, Überlingen, 1, 57, 60, 90, 156, 161, 75, 125-143
   Lockerbie, 73-74, 120
   Paris, Concorde, 109, 120
   Ramstein, 109
   Sioux City, 74-75, 81
   TWA Flight 800, 73, 94, 106
Aftermath, 19, 24, 72-73, 93-99, 105-106, 173, 176-177
Air Navigation Service Provider ANSP, 5-11, 57, 75-76, 81, 86-87, 90-91, 142, 157-158, 166-167
Air Traffic Control – Airport CISM Team ATC-AP, 76, 122-123
Air Traffic Control ATC, 5-11, 81-91, 125-143, 153-170
Air Traffic Controller ATCO, 62, 74-77, 81-91, 125-143, 153-170
Air Traffic Management ATM, 5-11, 81-82, 91
Airline, 93-107, 111, 124
Airline Flight-attendants Association AFA, 71, 100
Airline Pilots Association ALPA, 70-71, 76-77, 79, 106
Airport, 1, 3, 6, 10, 14, 19, 54, 59, 69, 76-77, 85, 109-124
Altruism, 66
American Red Cross, 72-73, 85, 93, 96-97, 99
Arousal, 26, 30, 51, 105
Association Professional Flight Attendants APLA, 100-101
Aviation psychology, 67-68
Avoidance, 30, 130, 156

Battery Model, 44-46, 157-158
Brain physiology, 30, 49-51, 83
Business benefits, 5-11, 153-170

Catharsis, 54, 60, 103
Close calls, 14, 25, 37
Cognitive behavior therapy CBT, 32
Communication, 5, 8-9, 14, 23, 42, 56, 62, 68, 76, 82, 87, 94, 98, 100, 111, 113, 123, 127-130, 155, 158-162, 168, 171-172
Community support, 38, 51, 55, 57, 61, 99
Confidentiality, 87, 10, 103, 105-106, 114-115, 127, 149
Coping strategy, 45-46, 53, 55-56, 58, 60, 130, 136, 139
Corporate recovery, 98, 100
Cost-benefit-analysis, 153-170
Crew resource management CRM, 68
Crisis intervention contacts, 20-21
Crisis intervention principles, 19-20, 150
   1. Proximity, 19-20, 23, 53
   2. Immediacy, 20, 53
   3. Expectancy, 20, 53
   4. Brevity, 20
   5. Simplicity, 20
   6. Innovative, 20
   7. Practical, 20
Crisis Intervention Team CIT, 18-19, 55, 68, 74, 76-77, 87, 90, 119, 122, 126-127, 141, 150
Crisis Management Briefing CMB, 18, 37-39, 55, 57, 171
Crisis reaction/response, 14-20, 26, 65, 71-72, 93-95, 98, 101, 148, 173, 175-176
Critical incident CI, definition, 43
Critical Incident Response Program CIRP, 18-19, 22, 30, 34, 70-71, 74, 77, 79
Critical Incident Stress CIS, 47-49
Critical Incident Stress Debriefing CISD, 18, 55, 59, 63, 93-94, 98, 100-102, 125-143, 148, 171-172
Criteria, 100-101
   Seven phases, 59, 102-103, 129-141
   1. Introduction, 59, 102-103, 129-130

2. Fact, 59, 103, 131-132
3. Thought, 59, 103, 132-134
4. Reaction, 59, 103, 134-136
5. Symptom, 59, 103, 136-138
6. Teaching, 59, 103-104, 138-139
7. Re-entry, 59, 104, 140-141
Critical Incident Stress Management CISM
  History in aviation, 67-79
  Tools, 18
Critical Incident Stress Reactions CIR, 8, 26, 29, 43-44, 47, 49-50, 52, 55, 82-83, 85, 89-90, 125, 136, 160
Culture, 87, 90, 93, 99, 111, 142, 156, 158-159, 168, 170
Customers, 36, 38, 110, 118

Defusing, 57-58, 100-102
  Three phases
  1. Introduction, 58, 101-102
  2. Exploration, 58, 101-102
  3. Information, 58, 101-102
Déjà-vu sensation, 50, 52
Demobilization, 18, 39, 55, 57, 98-99, 171
Desensitization, 32
Deutsche Flugsicherung (German ANSP) DFS, 1, 5-11, 57, 71, 75, 81
Disaster, 13-15, 25, 29, 38-42, 46, 51, 54, 57, 61, 63, 66-68, 72-75, 77-79, 90, 93-99, 101, 105-106, 109, 124, 134, 145, 148-150, 173, 175-176
Distress, 13-21, 24-25, 30, 32, 34, 36, 50-51, 57, 66, 69, 72, 74, 100, 103-104, 145

Emergency level, 96
Emergency Response and Information Center ERIC, 122
Employee Assistance Program/Professional EAP, 54, 68, 75, 96, 98-101, 173, 175
Eurocontrol, 2, 75, 77, 81, 91, 142-143, 153
Eu-stress, 50
Evaluation, 2-3, 8-9, 149, 153-170
  Mental health, 30
  Psychological, 30, 68, 136
Exposure, 24, 26, 29-30, 104-105, 127, 145, 150
Eye Movement Desensitization and Reprocessing EMDR, 33

Family support, 18, 37-39, 55, 60-61, 68, 70, 73, 97-99
Federal Aviation Administration (of the USA) FAA, 69-70, 75, 106
Fight/flight response, 49
First aid, emotional, psychological, 22, 23-24, 34-36, 40, 42, 54, 66-67, 73, 77-78, 83, 109
First responder, 93-98, 102, 106
Flashback, 52
Flight attendants, 14, 19, 36-37, 39, 71-73, 76-78, 100, 120
Frankfurt Airport Group FRAPORT, 1, 69, 76, 109-124

Greeters, 110-111, 114, 121
Group intervention, 18, 42, 55, 67, 75, 88, 126, 143, 146-147, 171-172, 175
Guilt, feelings of, 14, 29, 47, 54, 85-86, 88, 164

Hippocampus, 51
Homeostatic status, 50
Human Resource Performance Model HPM, 153, 156-157, 160, 163, 169
Hypothalamic Pituitary Adrenal axis HPA, 50

Implementation, 1-2, 7-8, 10, 81-82, 86, 91, 110, 117-118, 143, 153, 155, 159, 167
Instructors, 69, 72, 75, 149-150
Intangible, 2, 153, 156, 167
International Critical Incident Stress Foundation ICISF, 1, 10, 55, 70-76, 81, 87, 125-126, 142, 146
  European Office, 10, 76
International Federation of Air Traffic Controller Associations IFATCA, 75, 91, 142
Intrusive symptoms, 30, 50, 52, 104
Investigation Testimony, 105, 130

Limbic system, 49-50, 83

Manager, 5-11, 19, 55, 57, 69, 72-73, 85-90, 95-97, 109, 113, 115, 122, 124, 133, 142, 153, 155-156, 169

Managing supervisor, 5, 8, 49, 55, 67-68, 86, 89-90, 97, 118, 121, 139, 157-169
Media, 14, 38, 84, 97, 109, 110, 115, 130, 132
Meeters, 110-111, 114, 121
Mental Health Professional MHP, 15, 19, 21, 23, 30, 32, 37, 58-60, 68, 72-73, 75, 87-89, 93, 95-96, 99, 101-106, 120, 125, 127, 149
Mitigation, 9, 17, 22-24, 28, 36, 93-94, 102
Mobilization, 24, 36, 57, 94, 97

National Transportation Safety Board NTSB, 70, 97-99, 105-106
Natural disaster, 13, 41, 46, 54, 61, 78, 90
 Asian Islands, tsunami, 13, 90, 113-114, 121, 145
New Orleans, hurricane Katrina, 145
Neurobiological, 43-52
Nightmare, 30, 47, 52
Normalization, 24, 56, 59, 61-62, 139, 163

One-on-one crisis intervention, 22, 55-57, 100-101, 146, see also SAFER-R
On-scene support services, 55
Organizational support, 61, 90, 155

Parasympathetic, 50
Peer, 1-3, 7-8, 19, 44, 46, 49, 52, 57-60, 68, 81, 101, 118, 120, 121, 125-143, 154-170
 Certification, 75, 81, 86, 88, 90, 126, 142, 148, 150, 175
 Competencies, 145-146
 Conference, 10, 90, 169
 Election, 87
 Model, 44, 52, 62-63, 86, 88-89, history of model, 71-76
 Qualification, 52, 58, 85-86, 89, 91, 101-102, 127
 Role, 61, 63, 101, 128, 130, 132
 Selection, 87, 127
 Supervision, 85, 88-89, 155
 Training, 10, 15, 55-56, 61, 63, 87-88, 91, 145-151
Personal crisis, 36
Pharmacotherapy, 32

Pilot, 14, 19, 32, 36, 48, 62, 67-73, 76-77, 83, 85, 103, 106, 123, 141, 160, 162
Post Traumatic Stress Disorder PTSD, 22, 28-33, 43, 50-53, 71, 84-85, 104, 119, 147, 156, 171-173
Practitioner, 1, 146-149
Pre-incident, 18, 55-56, 93-95
Preparedness, 93-95
Presenteeism, 161-170
Psychogenic amnesia, 105
Psychological support, 68, 93-94, 98, 119
Psychophysical status, 44, 46-47
Psychotherapy
 Contrasted to CISM, 17, 21-24, 53-54, 104, 130, 173
 Referral, 22, 136-137

Recovery/reintegration, 93-94, 98, 99, 106
Referral, 18, 21-22, 25, 34, 39, 55-56, 61, 85, 136-137, 147, 172-173
Relaxation, 32, 50, 113
Religion, 28, 66, 98, 111-112, 116
Response level, 35-39
Return on investment ROI, 154
Rubicon model, 157

SAFER-R, 24-25, 36, 55-57, 85, 90, 171
 1. Stabilize, 24-25, 56
 2. Acknowledge, 24-25, 56
 3. Facilitate, 24-25, 56
 4. Encourage, 24-25, 56
 5. Recovery, 25, 56
 6. Referral, 25, 56
Screening, 67-68
Stabilization, 98, 100, see also SAFER-R
Standardization, 69-70, 72, 74, 146, 148-149, 169
Strain, 82, 133-134, 139
Strategic approach, 2-3, 13-40, 146-149, 153, 156-157, 160, 163, 166-170, 172, 175
Strategic formula, 35, 46
 1. Target, 35-39
 2. Type, 35-39
 3. Timing, 35-39
 4. Theme, 35-39
 5. Team, 35-39

Stress reaction, 8, 25-29, 32, 43-44, 46-55, 58, 62, 71, 82-85, 88-90, 125, 136, 160
  1. Common, 26
  2. Moderate, 26
  3. Excessive, 27
  4. Extreme, 27
  5. Dangerous, 27
Suicide, 25, 27-28, 52, 67, 112, 118-119, 121
  1. Prevention, 146-147, 172, 175
  2. Intervention, 146-147, 151, 172, 175
  3. Postvention, 146-147, 172, 175
Survivor, 41, 53, 68, 72, 96, 145
Sympathetic, 50-51

Team Resource Management TRM, 6-7, 90, 161, 164, 178
Terrorism, 173-175
  September 11th, 39, 75, 145
Thought Field Therapy TFT, 174
Trainer, 146, 149-150, 154
Trauma, 19, 22, 26, 28-35, 49-59, 71, 73-74, 94-96, 102-106, 109, 113, 117, 119-120, 127, 135, 145-150, 171-175
Trauma Incident Reduction TIR, 33
Trauma response team, 73

Victim, 13-14, 19, 29, 51, 53, 72-74, 96, 98, 113-114, 119-120, 142

World Health Organization WHO, 51